FLUID POWER TECHNOLOGY

[A Text Book For Degree Diploma in Mechanical Engineering and Practicing Engineers in India and Abroad]

[A Text-Book for Engineering Services. Competitive Examinations, U.P.S.C. Engineering Service Examinations, Civil Services, ICS Examinations and AMIE Examinations.]

By

Prof. Ram. S. Srivatsa

BE. MIE., Formerly Chief Engineer, Southern Industrials; Joint Director, Govt. Of India; Design Engineer, Sundaram Clayton Ltd & presently, Consultant, Fluid power Technologies. Bangalore, *Karnatka*

STANDARD BOOK HOUSE

Unit of: **RAJSONS PUBLICATION PVT. LTD.**

1705-A, Nai Sarak, P.B. No. 1074, Delhi-110006
Ph.: +91-(011)-23265506 • Fax: +91-(011)-23250212, 43551185
Show Room: 4262/3, First Lane, G-Floor, Gali Punjabian, Ansari Road, Darya Ganj, New Delhi-110002 Ph.: +91-(011)-43551085, 43551185
E-mail: sbh10@hotmail.com • www.standardbookhouse.com

Published by:
RAJINDER KUMAR JAIN
Standard Book House
Unit of: Rajsons Publications Pvt. Ltd.
1705-A, Nai Sarak, Delhi - 110006
Post Box: 1074
Ph.: +91-(011)-23265506 Fax: +91-(011)-23250212

Showroom:
4262/3, First Lane, G-Floor, Gali Punjabian
Ansari Road, Darya Ganj
New Delhi-110002
Ph.: +91-(011)-43551085, +91-(011)-43551185
E-mail: sbhl0@ hotmail.com
Web: www.standardbookhouse.com

Second Edition : 2015

Price: **₹ 220.00**

ISBN: 978-81-89401-28-3

Typeset by:
Amandeep Singh
Graphic Era (an organisation of Book designing)

Printed by:
R.K. Print Media Company, New Delhi–110039

ABOUT THE AUTHOR

Prof. Ram S. Srivatsa is a graduate in mechanical engineering from the University of Bangalore. He served in HAL, Bangalore, Sundram-Clayton Ltd, Madras and Prototype Development and Training center, Madras (an Indo-Danish collaboration company.) He resigned as Joint Director from Prototype Development and Training center to take up an assignment with Hohenforst Inc. USA. He returned to India in 1995 and served as a consultant to Gordan Woodroffe Ltd, Madras and Auro-Electronics Ltd, Pondicherry. Widely travelled both in India and abroad he worked in Egypt on a Unido Project and in Russia at the request of an American company.

All through his career spanning nearly 25 years, he has been practising hard-core engineering and his field of specialisation has been design of SPM's with particular reference to Hydraulic & Pneumatic machines. He has independently designed and developed both as an employee and as an employer nearly 20-25 Hydraulic machines from concept to prototyping and proving. In addition, he has rebuilt/reconditioned an equal number of Hydraulic & Pneumatic machines.

He has taught M. Tech. students of Anna University as a guest faculty at the request of Central Leather Research Institute, Madras; the functions of leather and foot wear machinery, which are mostly either Hydraulic or Pneumatic machines. He was a lecture in machine design at Hindustan Institute of Engineering Technology, Madras and at Datamatics Ltd, Madras.

He is currently practising as a consultant in Fluid Power Technology.

PREFACE

Fluid Power Technology is a term that encompasses both oil hydraulics and pneumatics. Oil hydraulics can further be classified into industrial hydraulics and mobile hydraulics. The hydraulics pertaining to machine tools, hydraulic presses etc. are dealt with under industrial Hydraulics. The hydraulics pertaining to construction machinery, lifting equipments, earth moving and agricultural equipments are dealt with under mobile hydraulics. The discussions in this book are confined to basic Industrial Hydraulics. Advanced Industrial Hydraulics using servomotors and servo valves are therefore beyond the scope of this book.

What is Fluid Power? Most engineering students just out of college and many practicing engineers would define it as the power of the impinging water whose potential energy is converted into kinetic energy, causing it to drive a prime mover, such as a Pelton wheel. Although, this definition is true, this indeed is not the subject matter of this book. Engineering students just getting out of the university know of, only, water hydraulics, a subject that deals with hydro-electric power generation and transmission. The forces involved here are, hydrodynamic. Fluid power technology on the other hand deals with hydrostatic forces. This technology unlike water hydraulics does not depend upon an existing potential energy source. A positive displacement pump, will impart the energy to a confined fluid. This energy is used to generate force, transmit and control motion. The extension and retraction of the aircraft landing gears is achieved through the adaptation of this technology. So also, the to and fro motion of the bed of a grinding machine.

This technology has developed in a relatively short span so rapidly that there is hardly any industry that does not make use of this technology today in one form or the other. It is as much an integral part of any industrial activity as electronics.

The worldwide market for this industry is around $15 billion. Of this the US market share is 46% followed by Japan (19%), Germany (18%) and UK (6%). The rest of the European world shares the balance. India's share at 0.5% is less than that of Denmark. It appears that the technological advancement of any country is directly proportional to its consumption and distribution of fluid power products.

The fact that such a vital subject as Fluid power technology is even today not a part of the mechanical engineering curriculum in engineering colleges in India has never ceased to amaze me. To that extent I consider mechanical engineers who are product of our universities as half-baked, since most of them have never received even diploma level exposure to this subject.

The 'state of the art' of this technology in India is even today confined to those few, who have acquired this technology over the years out of shear practice.

This in fact should adversely affect our technological development. It is well known that the competitive performance and cost of a product is determined at the design stage itself to the extent of eighty percent. It is important therefore that the engineer has a sound understanding of all the available technologies to enable him to choose the most appropriate one in terms of cost competitiveness and performance criteria. The product designers who have had no exposure to this vital technology are therefore seriously handicapped while taking decisions during the critical concept stage of the design process.

Trained manpower therefore, be it engineers or technicians, is the need of the hour. Only a few institutions in India today offer fluid power engineering in their curriculum as electives. Due to the persistent efforts of **Fluid Power Society of India,** the Vishweshwariah Technological University of Karnataka, I understand, has come forward to introduce fluid power engineering as electives in undergraduate and postgraduate levels. This is a welcome trend. Lack of authoritative textbooks on this subject, is another area that needs to be tackled. Here again, it is the software and not the hardware that is the priority. Actuators, pumps and all kinds of control valves including cartridge, proportional valves and seals are available across the shelf. What needs to be taught is how this hardware should be put together in a circuit to obtain the desired results.

It is against this scenario that I decided to bring out this book. Obviously, my own academic qualification is not the basis, although a degree/diploma level education in mechanical engineering is required to understand the basic concepts of fluid power technology. It is my vast exposure to this subject as a practicing engineer both in India and abroad, in more than one type of industry and the nearly 20–25 different types of hydraulic machines that I designed from concept to prototyping and testing, is what in fact has encouraged me to bring out this book. Some of these machines have been included as exhibits in this book. My more than 25 years of association with **Fluid Power Society of India,** the only registered body that has been propagating the growth of this technology, has further induced me.

This book has been so fashioned as to kindle the interests of both the "high and the low" in the field of fluid power technology and will be published in three volumes. Volume one covers basic Industrial Hydraulics. Servo-actuators, servo-valves cartridge valves and proportional valves therefore form a part of volume-two under the caption 'Advanced Industrial Hydraulics and Pneumatics'. Volume three deals with mobile hydraulics.

Jan. 2009 **RAM S. SRIVATSA**

INTRODUCTION

The following are the three motion control technologies available as options to machine designers today.

1. Electro-Mechanical drives
2. Electro-Pneumatic drives
3. Electro-Hydraulic drives.

Choice of a suitable drive depends basically on operational criteria and cost factors.

The question often asked is why go for Oil Hydraulic systems when well known and reliable Mechanical, Electrical and Pneumatic systems are at our disposal. It is true, the steam engine, the internal combustion engine, the electric motor and the water turbines have all performed well as prime movers. However, they are not as compact or as efficient in directing the motive power to do useful work, as compared to hydraulic systems because, conventional gears and drive shafts impose a dimensional limitation on the designer. Also, mechanical systems require precision engineering and hence are not suitable for low cost innovations.

Electro-mechanical drives use complicated mechanisms such as ball screws, gears etc., to achieve linear motion or the complicated but well known piston connecting rod and crank shaft mechanism to achieve rotary motion. Good positional accuracy up to 0.025 mm is possible with ball screws and servomotors when the speeds are low and the loads are less than 1000 kg.

Electro-pneumatic drives are good for very light loads and high speeds but positional accuracies are difficult to achieve due to compressibility of air.

If linear or rotary motion is required with high speeds and heavy loads then the only choice is the electro Hydraulic drive. They are preferred even when the speeds are lower, as accurate position or force control is possible. The desired motion can be achieved with a very simple mechanism such as a hydraulic cylinder and piston rod assembly or a hydraulic motor.

Briefly, the following are the advantages of the oil hydraulic system over mechanical, electrical and pneumatic systems.

- Unlike in mechanical systems, the force is exerted in all directions.
- Oil Hydraulic systems are easier to comprehend and design since the laws governing the mechanics of fluids are simpler than the mechanics of solids, vapours or electricity.
- Oil hydraulic drives are very compact and can attain a weight to power ratio 0.5 to 0.8 kg/kW, values almost impossible to attain in electrical or mechanical systems.
- Gigantic forces of the order of 5000 tonnes can be developed with relatively low input power and fewer working parts.
- Hydraulic systems can be operated under continuous, intermittent, reverse or even stalled conditions. Dynamic braking is possible.
- Hydraulic systems have high torque to inertia ratios. So all the torque produced can be applied to the load since the torque required to overcome its own inertia is negligible.
- Both force and motion can be easily transmitted and precisely controlled.
- Oil Hydraulics and Pneumatic systems are less hazardous than mechanical or electrical systems.
- If an electrical or mechanical power source is disconnected, all of the potential energy in the system disappears. Whereas in oil Hydraulic and Pneumatics systems the potential energy is available even after the prime mover is disconnected. This is particularly advantageous during sudden power failures.
- Hydraulic Oil is a good conductor of heat. It also acts as a lubricant thus assuring long component life.
- The availability of both linear and rotary actuators makes oil Hydraulic systems highly flexible.

Contents

1

FUNDAMENTALS

'Pressure exerted on a confined fluid is transmitted equally in all directions and acts with equal force on equal areas'. This principle enunciated by the French scientist, *James Pascal* sometime during the 17th century was lying dormant for more than five decades till *Joseph Bramah* appreciated the significance of this principle and invented the hydraulic press in 1795. Pascal's principle should explain why a glass bottle filled with water up to the brim should break, when the stopper is forced in.

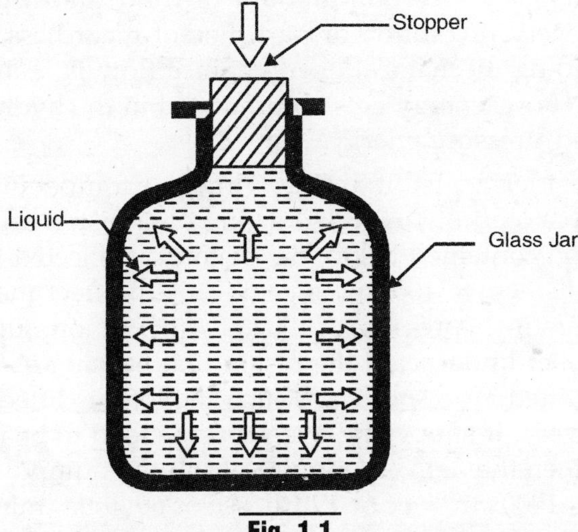

Fig. 1.1

1.1 WHAT IS PRESSURE?

The force acting on a unit area is defined as pressure. The area that we consider here is the area at right angles to the direction of the force. The unit of measurement of pressure is kgf/sq-cm in the MKS system or lbf/sq-inch in the FPS system or Newtons/sq-metre (Pascals) in the SI system. Whatever the units, the numerator always represents *force* and the denominator represents *area*. So pressure is *force/unit area.*

Pressure by itself has only a theoretical value and should be of interest only to the system designer. Force on the other hand has a practical value. It represents the muscle power. Its unit of measurement is kgf, lbf, or Newtons. Hydraulic machines are not designed to develop a certain pressure but they are designed to develop a certain force. You cannot determine the muscle power of a hydraulic machine by looking at the pressure indicated on the pressure gauge since you are looking at merely a component of the force and not the force itself. Only when you multiply this pressure by the cylinder area you will know the output force for which the machine is designed or the capability of the machine.

1.2 OIL HYDRAULIC SYSTEM IN A NUTSHELL

Any hydraulic system consists of three main parts. The first part is the hydraulic pump and the second part is the actuator and the third part is the fluid medium. The actuator can be a hydraulic cylinder or a hydraulic motor depending upon whether a linear or a rotary motion is desired as output. Primary element, the pump, when driven by a prime mover is designed to displace a certain amount of fluid that it draws from the reservoir, for every revolution or part thereof regardless of resistance to flow by expending mechanical energy. Since flow is maintained in spite of resistance to flow, energy gets accumulated in the hydraulic fluid. This energy is called pressure energy.

Let us consider a piston and a cylinder arrangement as shown in Fig. 1.2. Due to a load of 10,000 kgf acting on a piston area of 100 sq-cm, the hydraulic oil contained below the piston is subjected to a pressure of 10,000 kgf/100 sq-cm = 100 kgf/sq-cm. This in effect means, if a pump capable of generating a pressure of 100 kgf/cm^2 or more supplies hydraulic oil to the chamber underneath the piston that has an area of 100 cm^2, the load could be lifted. The speed at which the load is lifted depends upon the pump delivery. If supposing, we want the load to be lifted at the rate of 10 cm/sec, then the delivery required from the pump would be 10 cm/sec $\times$ 100 cm^2 = 1000 cm^3/sec or 1 litre/sec or 60 liters/min. There are two ways to bring the load down.

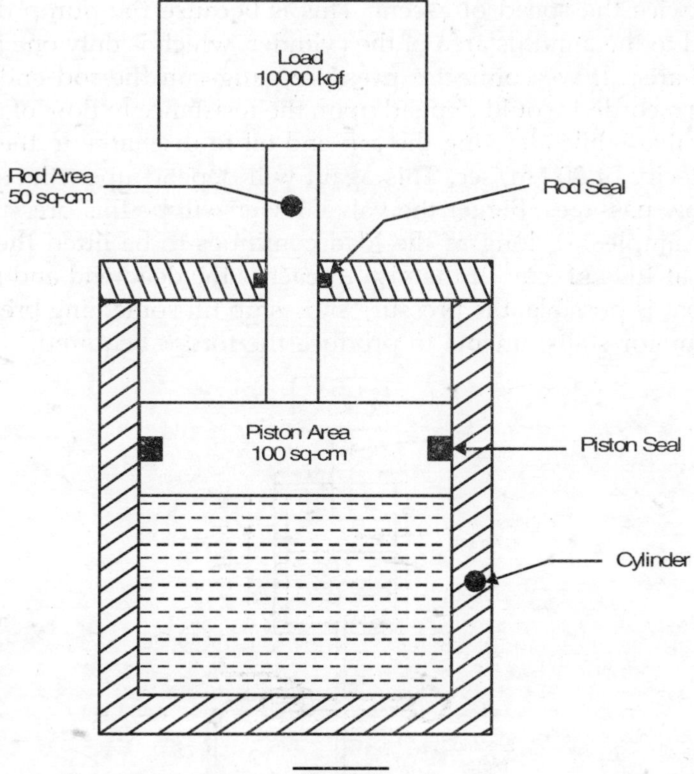

Fig. 1.2

If you open the valve (4), in (Fig. 1.3), the chamber is depressurized, since the fluid is exposed to tank line pressure and the load descends at a speed governed by the rate of discharge of oil to tank, which in turn depends on the area of the valve flow passages. It is interesting to note, the load descends upon opening valve (4) even when the pump is running because the pressure builds up only when there is resistance to flow.

The pressure gauge inserted in the line would show zero as all the oil is diverted back to tank.

By installing the valve (5) as shown in Fig. 1.4 we can control the up and down motion of the load simply by shifting the hand lever of valve (5). This changes the direction of flow of oil from the pump to the rod end of the cylinder while the cap end gets connected to tank. Note that the speed of descent depends on the pump delivery and the annulus area at the rod end. The speed of descent in this case would be

$$\frac{100 \text{ cm}^3/\text{sec}}{(100-50) \text{ cm}^2} = 20 \text{ cm/sec}$$

This is twice the speed of ascent. This is because the pump delivery is admitted to the annulus area of the cylinder, which is only one half of the cap end area. If we connect a pressure gauge on the rod end side, the pressure recorded would depend upon the resistance to flow of oil offered by the valve while allowing the cap end oil to discharge to the tank at a flow velocity of 20 cm/sec. This again will depend upon the size of the valve flow passages. Bigger the valve, lower will be this pressure. In the above examples, so long as the load continues to be lifted the pressure remains at 100 kgf/cm^2. If the piston reaches the dead end and no further movement is possible, the pressure shoots up till something breaks or the electric motor stalls, unable to produce the torque required.

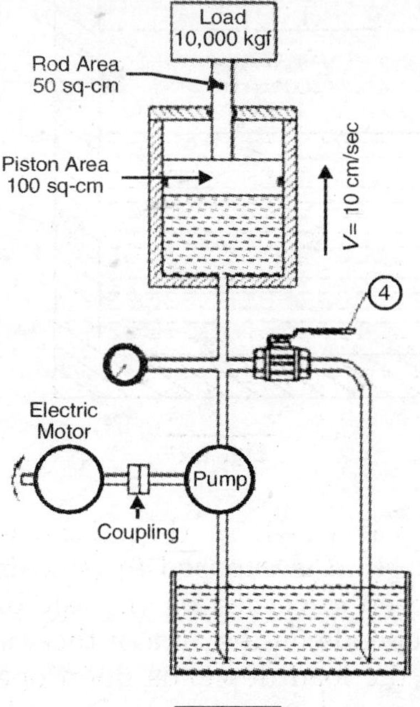

Fig. 1.3

In order to prevent such failures a valve, which acts like an electrical fuse is incorporated in the circuit, in the pressure line as shown in Fig. 1.4. This valve will ensure that the pressure in the chamber shall never exceed the set pressure, which in this case is 100 kgf/cm^2. At pressures higher that this, the valve opens and diverts the excess flow to tank.

We have just designed a simple hydraulic circuit to lift a load of 10,000 kgf at a speed of 10 cm/sec. We have chosen 100 kgf/sq-cm as the operating pressure and 60 lpm as the required delivery from the pump to accomplish this task on a cylinder area of 100 sq-cm. If the operating pressure is reduced, the cylinder area should correspondingly be increased, so that the force requirement is kept constant.

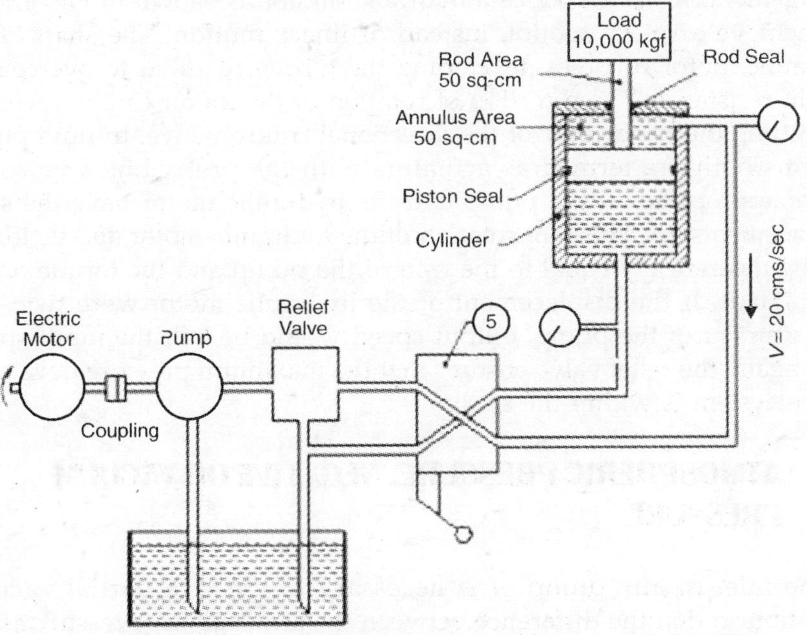
Load
10,000 kgf
Rod Area
50 sq-cm
Rod Seal
Annulus Area
50 sq-cm
Piston Seal
Cylinder
V = 20 cms/sec
Electric
Motor
Pump
Relief
Valve
5
Coupling

Fig. 1.4

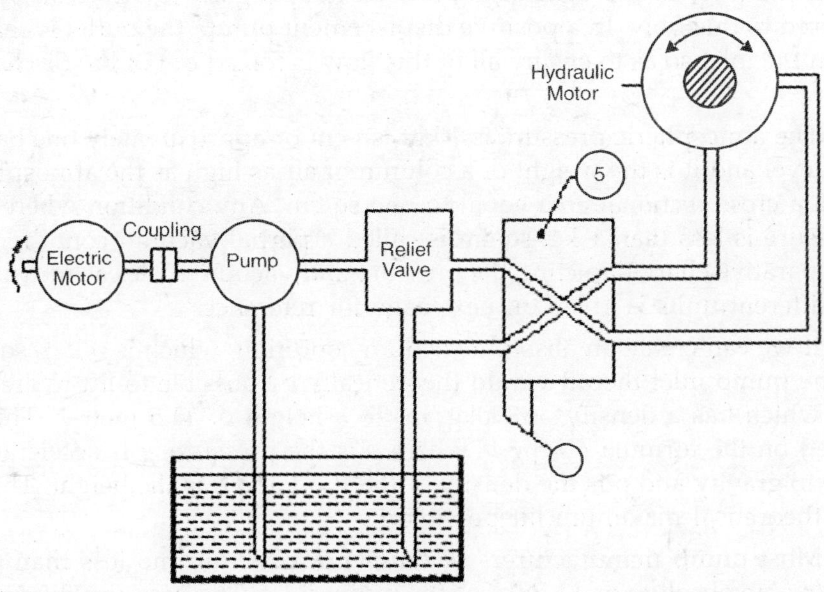
Hydraulic
Motor
5
Coupling
Electric
Motor
Pump
Relief
Valve

Fig. 1.5

In the above example, in place of the hydraulic cylinder and piston arrangement, if we introduce a hydraulic motor as shown in Fig. 1.5, we can achieve a rotary motion instead of linear motion. The shaft of the hydraulic motor revolves developing the torque required to overcome a certain resistance. The direction of rotation of the motor can be reversed, by shifting the hand lever of the directional control valve. In fluid power systems both are termed as actuators with the prefix linear or rotary where appropriate. If the pump and the hydraulic motor have the same displacements, the speed of rotation of the hydraulic motor and its torque will be theoretically equal to the rpm of the pump and the torque on the pump shaft. If the displacement of the hydraulic motor were twice the displacement of the pump, output speed would be half the input speed. Here again, the relief valve ensures that the maximum pressure developed by the system is within the set limits.

1.3 ATMOSPHERIC PRESSURE, NEGATIVE OR VACUUM PRESSURE

At the inlet of any pump, it is necessary to create a partial vacuum condition so that the difference between the atmospheric pressure acting on the fluid in the reservoir and the negative pressure at the pump inlet creates a flow due to pressure difference. The purpose of any pump is to create this condition at the inlet and transfer ideally all of this flow from the inlet to the discharge port, without regard to any resistances offered to this flow. In a positive displacement pump, the outlet is sealed from the inlet so as to ensure all of this flow is forced out of the discharge port.

The atmospheric pressure is 1 kgf/sq-cm or approximately one bar at sea level and it is the weight of a column of air as high as the atmosphere with a cross sectional area equal to one sq cm. Any condition where the pressure is less than 1 kg/sq-cm is called a partial vacuum condition. A comparative chart, depicting the pressure and vacuum scales as measured in different units is given on next page for reference.

If we can create an absolute vacuum condition, which is 0 kgf/sq-cm at the pump inlet then it would theoretically be possible to lift hydraulic oil, which has a density of 890kg/m^3 to a height of 11.5 metres. This is based on the formula $p = \rho g\ h$, where p is the pressure, g is acceleration due to gravity and ρ is the density of the fluid and h is the height. This is the theoretical maximum lift possible.

Most pump manufacturers recommend a vacuum no less than 0.85 kgf/sq-cm absolute or 12 PSIA at the pump inlet. The pressure difference available to create flow is only 0.15 kgf/sq-cm which in effect means

Pressure and Vacuum Scales

	PSI Absolute	PSI Gauge	kgf/cm² Absolute	kgf/cm² Gauge	MM. of Hg Absolute	Inches of Hg. Absolute	Inches of Hg. Absolute	Ft. of Water
One Atmosphere	14.7	0	1.0	0	760	29.92	0	34
	10		0.7		500	20	10	22.33
	5		0.35		243	10	20	11.5
Perfect Vacuum	0		0		0	0	29.92	0

excessive lift must be avoided. The maximum lift based on this recommendation is therefore 11.5 metres × 0 .15 = 1.7 metres considering an oil density of 890 kg/cubic metre.

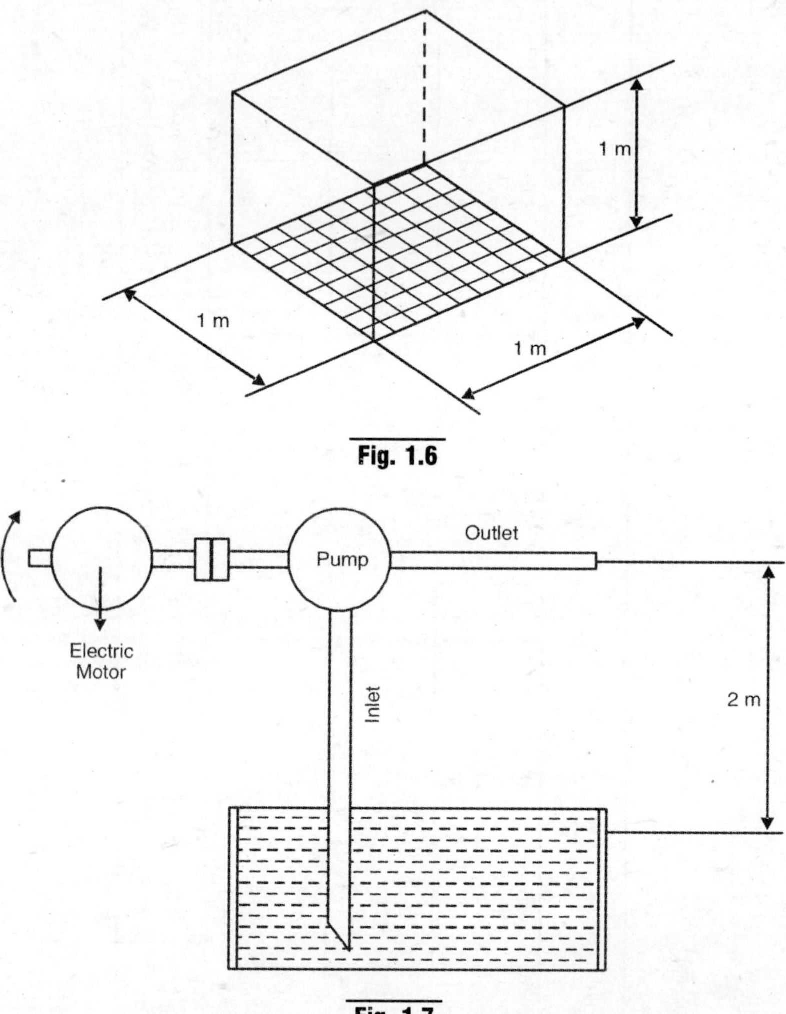

Fig. 1.6

Fig. 1.7

Excessive vacuum creates a condition favourable for vaporisation of oil leading to the formation of gas bubbles. These bubbles upon encountering positive pressure at the outlet, suddenly collapses with considerable force. This causes the so-called cavitations corrosion that damages the pumping elements, reducing its service life. This is, regardless of the fact that most hydraulic oils have good vapour pressure characteristics. If the inlet line fittings are not perfectly tight a somewhat similar condition is experienced since atmospheric air enters the inlet line and mixes with oil. This results in a noisy pump, erratic valve and actuator functions. It is therefore important to make sure the pump inlet line is placed as low as possible with respect to oil level in the tank and to ensure that all inlet line fittings are properly sealed and leak proof.

The problems associated with placing the pump above the tank can be all together eliminated if the pump is placed below the oil tank as shown in Fig. 1.8 or submerged in oil. The only problem with this kind of

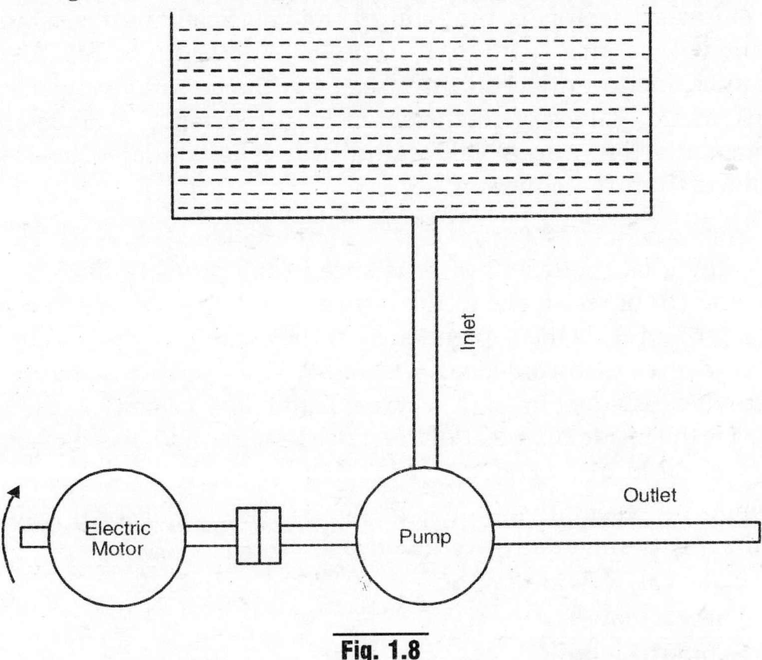

Fig. 1.8

an arrangement is that it is somewhat difficult to service the system during maintenance and repair.

1.4 FLOW THROUGH PIPES AND FITTINGS AND PRESSURE DROP

Hydraulic oil flowing through a pipe with a certain velocity encounters resistance to flow. If the flow passages are adequately sized based on the recommended fluid velocity, then this resistance is limited to pipe wall friction, turbulence due to sudden enlargement and sharp bends. This resistance to flow causes a pressure drop between two points and results in wasted power. A good design practice is to ensure that this wasted power is minimum.

Resistance to flow due to friction in the pipes cannot be avoided unless the walls are made entirely smooth. For pipelines of length less than

100 d where d is the inside diameter of the pipe, the frictional losses in pipes are never taken into account, as they constitute a small portion of the total energy input. The losses in valves and fittings is what is usually determined. The losses in valves and fittings, are proportional to the square of the velocity of the fluid that is, $\Delta p \alpha v^2/2g$. The loss of energy or the pressure drop is determined by the equation,

$$\Delta p = (K_0 \gamma v^2/2g) \text{ kgf/m}^2 \quad (1)$$

K_0, an empirical factor, is the sum of the coefficients of resistance for different valves and fittings, v is flow velocity in metres/sec, g is acceleration due to gravity in mts./sec^2, γ is the specific weight of oil in kg/cubic metre. Values of coefficient of local resistance, K as determined experimentally for various valves and fittings is tabulated below. K_0 is determined from the equation,

$$K_0 = [\, K_1 + (A_2/A_1)^2 K_2 \,] f \quad (2)$$

K_1 is the total coefficient of resistance in the pressure lines, K_2, is the total coefficient of resistance in the return lines. A_2 is the rod end area of the cylinder and A_1 is the cap end area of the cylinder. f is the correction factor for degree of turbulence $= (2000/R_e)^{0.25}$ and R_e is the Reynolds number which is equal to vd/n where v is the flow velocity in metres per second, d is the inside diameter of the pipe in metres and n is the kinematic viscosity of oil in m^2/second.

Certain representative values of the coefficients of local resistances determined experimentally are given below.

1. Gate valve (wide-open). 0.20
2. Large elbow 0.33
3. Standard elbow 0.90
4. Standard-T 1.80
5. Elbow 45º 0.43
6. Cross, (branch flow) 1.20
7. Return band 2.20
8. Ball check valve 4.00

Values of pressure drop Δp, due to local resistances obtained based on the above formulae are sufficiently accurate for most purposes. In case of very long and complicated pipe lines the energy loss must be calculated taking into account the head loss due to friction also, using Darcy's equation, in addition to energy loses due to local resistances. Then Bernoulli's and continuity equations may be applied to calculate the pressure drops.

No standard recommendations exist regarding permissible pressure drops in a circuit. They are difficult to control because much depends on how well the installation job is carried out and the number of hydraulic valves and fittings used to achieve a certain end result. The recommended flow velocity in pipelines is 1.5–2 metres per second for the suction lines. For the pressure lines the recommended flow velocities are any where between 3–7 metres per second depending upon the system pressure selected. Lower the system pressure, lower will be the recommended flow velocity in the pressure lines.

Based on this recommendation, the inside diameter of the pipe required to handle a certain flow is determined from the equation

$$d_0 = 4.6\sqrt{Q/v},$$

where d_0 is the inside diameter of the pipe in mm,
Q is the flow rate in liters/min and
v is the recommended velocity in metres/sec.

Most hydraulic component manufacturers would have given the pressure drop characteristics of their products at certain flows and this factor should be considered while evaluating the total resistance to flow and hence the pressure drop. A fair estimate of the resistance to flow offered by the system can be evaluated by measuring, the current drawn by the electric motor while driving the pump which is idling, that is, when the pump is simply circulating the oil back to tank through the valves. This current should be slightly more than the rated current for the motor at no load. This factor represents the efficiency/the energy loss and has to be balanced against the cost of larger pipes, fittings and better installations.

A straightforward equation that can be used for calculating the pressure drop along pipelines is given below.

$$\Delta P = (2 \times 10^{-5}) f \rho v^2 L/d \qquad (3)$$

Δp is the required pressure drop in bar (1bar ≈ 1 kgf/sq-cm), $f = 64/R_e$, L is the length of the tube in meters, ρ is the density of the liquid in kg/m^3, v is the flow velocity in meters/sec, and d is the inside diameter of the pipe in metres. The value (2×10^{-5}) is a constant for the units employed.

WORKED EXAMPLE

Problem. *Find the pressure drop over a length of 5 meters of hydraulic oil of specific gravity = 0.89 and kinematic viscosity = 25 cSt flowing @ 250 cm^3/sec through a tube of 10 mm diameter. Neglect loss through fittings.*

Solution. The density r of the liquid is (1000 kg/m^3× 0.89)= 890 kg/m^3, Kinematic viscosity n =25 cSt = 25 × 10^{-6} m^2 /sec, flow velocity v = volume flow rate in cubic metres/sec × flow passage area in sq.metres = (250 × 10^{-6} × 78.54 × 10^{-6})= 3.183 metres/sec. d = inside diameter of the pipe =

10^{-2} meters, $f = 64/R_e$ where R_e is given by vd/n. $R_e = (3.183 \times 10^{-2})/25 \times 10^{-6} = 1,273$. Since $R_e < 2000$ the flow through the pipe can be considered laminar. If R_e was more than 2000 then the flow is considered turbulent in which case the value of $f = (0.316/\sqrt[4]{R_e})$.

Substituting the above values in Eq. (3) we have $\Delta p = (2 \times 10^{-5}) f \rho v^2 L/d$ or $\Delta p = (2 \times 10^{-5})$ $[64/1,273 \times 890 \times (3.183)^2 \times 5/10^2] = 1.13$ bar.

1.5 MANOMETERS AND FLOW METERS

The manometer also called the pressure gauge is a very familiar component of the oil hydraulic system. Pressure gauges are required to evaluate the performance of fluid power components and can be helpful in troubleshooting. But their most important function is to help set the system relief valve to desired pressures to control the force exerted by a cylinder or the torque developed by a hydraulic motor.

Fig. 1.9. *Pressure gauges.*

Pressure gauges generally are of the Bourdon tube type formed into an arc or a spiral. Metal tube is bent to form an arc one end of which is closed and the other end is connected by means of tap (1) to the point where the pressure is measured. When the high-pressure fluid is admitted via valve (1) to the Bourdon tube it tends to straighten out. This actuates the linkage to the pointer gear (2) and moves the pointer to indicate the pressure on the caliberated dial. When the pressure is released the tube assumes its original position and the gauge reads zero. The gauges read zero at atmospheric pressure and are calibrated to read pressures both in pounds/sq. inch and in kgf/sq-cm or bar.

One bar is ≈ 1kgf/sq-cm. The atmospheric pressure is always excluded. For obtaining absolute pressures the atmospheric pressure which is

1 kgf/sq-cm. or 14.7 pounds/sq. inch must be added to the gauge pressure indicated.

To increase gauge life and reduce the effect of system pulsations and mechanical vibrations glycerin filled gauges are available. Glycerin, a stable viscous fluid dampens the vibrations of the Bourdon tube and the pointer to ensure positive gauge read out.

Pressure gauges are generally manufactured in 4 different dial sizes. 65 mm, 100 mm, 150 mm and 250 mm are standard sizes. There are models that can be mounted directly on the fluid line. Models, which are of three-hole panel mounting type, are also available.

They are available to read pressures up to 1400 kgf/sq-cm in different dial sizes. For low and medium pressure applications normal arc type phosphor bronze Bourdon tube would be sufficient. For very high pressures spirally wound tube is recommended.

The maximum dial reading recommended is no more than 65% of the dial range for constant pressures and 60% for variable pressures.

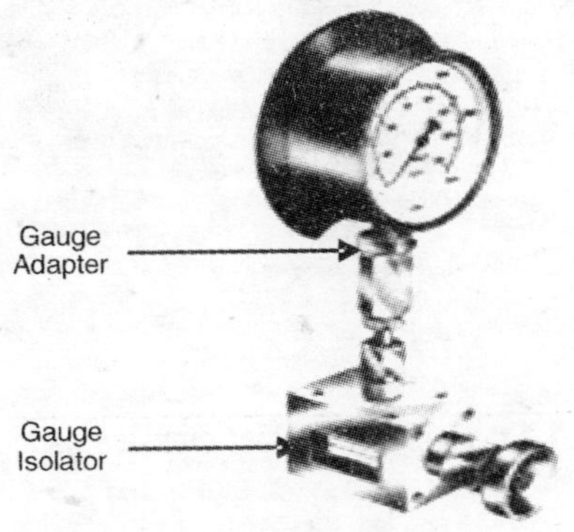

Fig. 1.10

While connecting the fluid lines pressure gauge adapters with copper or rubber seals must invariably be used for positive sealing at the entry port to the gauge. Push to read gauge isolators or shut-off valves if installed at the entry port to the gauge would not only enable gauge servicing without interrupting the system but also would help prolong the life of the gauge. Gauge isolators normally vent the gauge to tank when gauge reading is not required.

Magnetic electrical contact pressure gauges not only read system pressure but also help control the pressure at the set value by switching 'off' or 'on' solenoid operated directional control valves thus eliminating the need for a separate switching control for these valves.

Flow meters for measuring the pump flow rate, although not mandatory like the pressure gauges on all hydraulic systems is a vital instrument that helps evaluate the condition of the positive displacement pumps. The condition of positive displacement pumps cannot be evaluated merely by its capability to build up the rated pressure. It is equally important to determine whether the rated flow is maintained at the rated pressure levels.

Fig. 1.11. *Electrical contact pressure gauge.*

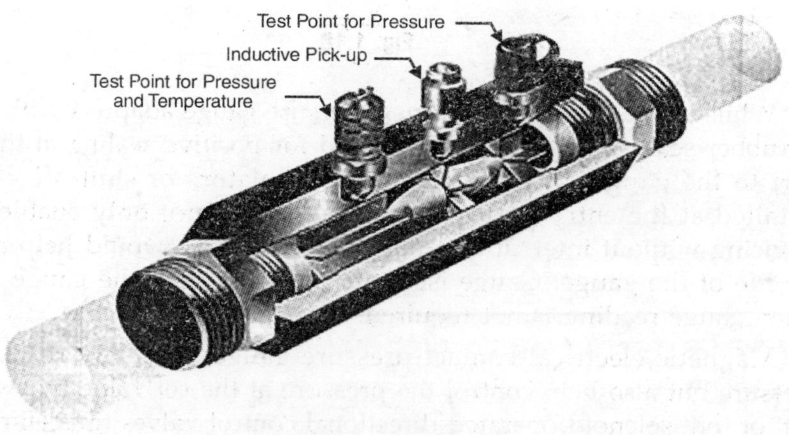

Fig. 1.12. *Flow meter.*

A typical flow-measuring device consists of an axial flow turbine blade wheel driven by the flow stream rotating in proportion to the mean flow velocity. A non-contacting inductive pick up generates a pulse each time its electro-magnetic field is interrupted by the rotating blades of the turbine.

These pulses are directly converted into a flow measurement in litres per minute by the associated electronic instrumentation. Additional test points for measurement of temperature and pressure are integrated in to the flow meters.

2

POWER INPUT DEVICES—HYDRAULIC PUMPS

Pumps mainly are of two kinds, Hydrodynamic and Hydrostatic. Hydrodynamic pumps are merely fluid transfer pumps. They are basically meant to transfer fluids from one location to another overcoming very little resistance. The weight of the fluid and pipe wall friction is about all that they have to overcome. While they can be designed for large flows their capacity to overcome any resistance to flow is very limited due to internal leakage. They are therefore termed *non-positive displacement pumps* and are not used in oil hydraulic systems and therefore, not considered here.

Hydrostatic pumps are also called the *positive displacement pumps*. This means the pump discharge will remain almost constant regardless of resistance to flow. Such pumps are ideal for oil hydraulic systems where large forces have to be developed by application of high pressures on small areas. These pumps can deliver hydraulic oil @ 3–3000 LPM and can withstand pressures from 35–1000 kgf/sq-cm. For continuous duty however, the available pumps have a range between 35–700 kgf/sq-cm.

2.1 PUMP CHARACTERISTICS

Industrial hydraulic systems can be classified as high-pressure systems (200–400 kgf/sq-cm), medium pressure systems (100–200 kgf/sq-cm) and low-pressure systems (50–100 kgf/sq-cm). The overall cost of low-pressure systems is higher than high-pressure systems. The low-pressure system occupies more floor space since the weight and size of actuators, oil tank, and other components are large. High-pressure systems on the other hand are compact and on the whole, are less expensive. The modern trend is to go for high-pressure systems, since the problems hither to associated with it, such as availability of high-pressure pumps and seals are, largely overcome.

Pump pressure rating is one of the major considerations for selecting a pump, as that would determine whether the pump could do the intended job of resisting a force. Equally important is its flow rating at the operating speed. This will determine whether the actuator can be moved at the desired speed. Other, less important features are, the noise generated at the operating speed, volumetric efficiency, cost, reliability, maximum and minimum operating speeds and compatibility with the fluid medium. In positive displacement pumps, the discharge remains constant despite load resistance. This means in a Cartesian co-ordinate system with flow (Q) on the Y-axis and pressure (P) on the X-axis, the function $Q = f(p)$ is a straight line. This however is the ideal condition. In actual conditions, due to inevitable volumetric losses, this characteristic is inclined at an angle to the theoretical. Higher the volumetric losses, greater will be this angle of inclination.

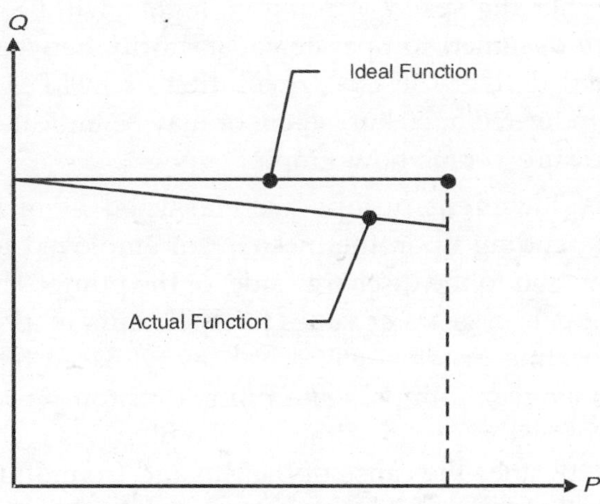

Fig. 2.1

This in effect is indicative of the volumetric efficiency of the pump. The discharge from the pump decreases to that extent at operating pressures. This factor should be taken into account along with the pressure drop across the valves and fittings while selecting and sizing the pump. Volumetric efficiency basically depends on the working clearances both radial and axial, between the pumping element and the housing. As this clearance increases the volumetric efficiency decreases. External gear pumps have the lowest volumetric efficiency, about 80 percent as compared to the piston pump that has an efficiency of about 95 percent. Dissolved air, which is always present in the hydraulic oil, also contributes to the inefficiency of the pump since the suction chamber is only partly filled with oil resulting in reduced discharge. Foaming of hydraulic oil and formation of air oil emulsion are the indicative factors. The problem of dissolved air appears to affect axial piston pumps more than gear or vane pumps probably due to the dead space in their working chambers.

The input power required to drive the pump is a function of the delivery (Q), and the pressure required to overcome load resistance (p) including the pressure drop. The volumetric and the mechanical efficiencies of the pump, η_v and η_m respectively are combined and approximated to one value based on the characteristics of the pump selected.

The input power HP required to drive the pump is given by, HP $= p \times Q/(600 \times 0.746 \times \eta)$. P is the system pressure in kgf/sq-cm, Q is the discharge in liters per minute, η is the combined efficiency of the pump selected. Pump discharge depends on the speed of rotation or RPM of the pump shaft. Higher the speed of rotation, higher will be the discharge. Most pumps are designed to operate satisfactorily between 600 to 4000 RPM. Their rated delivery in cm^3/revolution or in LPM at 1200/1400 RPM is usually indicated at 70 kgf/sq-cm or may be indicated for different speeds on a pressure versus flow graph.

Positive displacement pumps are classified as either rotary or reciprocating depending upon the mechanism employed to displace the fluid from the suction to the discharge side, of the pump. This mechanism can be a rotating pair of gears or vanes freely floating in a rotating disc or plungers reciprocating inside a cylindrical cavity. Based on this, they are designated also as gear pumps, vane pumps, piston pumps and screw pumps. In case of piston pumps although the plungers reciprocate inside a cylindrical cavity since the cause of their to and fro motion is a rotating element, they cannot be classified strictly as reciprocating pumps except for the hand operated versions.

2.2 HAND OPERATED PISTON PUMP

Perhaps the most elementary of all positive displacement pumps, is the hand pump consisting of a plunger which is free to slide up or down in a cylindrical cavity. The clearance between the piston and the cylinder is closely controlled so as to prevent air entering the chamber. The plunger is hinged to a hand lever which when moved up or down, extends or retracts the plunger in its cavity. The principle of positive-displacement, is well illustrated in this pump. As the plunger extends, a partial vacuum is created due to which oil in the reservoir enters the cylindrical cavity below the plunger through the inlet check valve. The volume of fluid drawn into the chamber is the product of the cross sectional area of the plunger and its displacement. When the hand lever is pushed down the plunger retracts forcing all of this oil to exit through the outlet check valve.

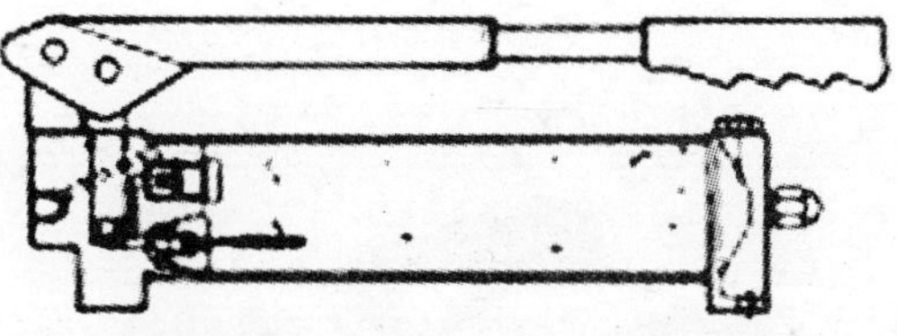

Fig. 2.6. *Hand operated piston pump.*

The pump discharge depends on its pressure rating. Higher the pressure rating, lower the discharge. The limiting factor is the hand force required at the end of the hand lever at the rated pressure and flow. This is limited to about 35 kg. Depending upon their construction, these pumps will deliver between 3 cc per stroke to 50 cc per stroke and can build up pressures up to 700 kgf/sq-cm. A double acting pump, which delivers fluid even during the upward stroke, can deliver higher flows.

Pumps, which automatically change over from high discharge at low pressure to low discharge at high pressure, are also available. Such pumps with a pressure rating of 700 kgf/sq-cm, can deliver up to 50 cc per stroke at pressures below 30 kgf/sq-cm and automatically change over to a reduced flow of 3 cc per stroke at pressures beyond. These pumps are useful where the actuator has to travel a considerable distance before the load resistance is encountered. They can accomplish the same job in about half the time that it takes for a single stage pump.

The number of strokes required to accomplish a certain task can be determined from the rated delivery of the pump and the cross sectional area of the actuator involved. For instance, if the pump discharge is 4 cc/stroke and the actuator area is 12 sq-cm then the actuator would move a distance equal to 4/12 cm or 0.3 cm in one stroke. This when divided by the total distance the piston has to travel, will determine the number of strokes.

2.3 GEAR PUMPS

These are of two types. External gear pump and internal gear pump. Among the two the external gear pump is more popular. This may be due to the fact that internal gear pumps are more complicated to manufacture and therefore are more expensive without an appreciable difference in performance. External gear pumps have fewer working parts and therefore are compact and simple in construction compared to other pumps. They are also the least expensive. They tolerate cavitation and are resistant to shock. These are recommended for low pressure, medium flow application.

Fig. 2.2. *External gear pump.*

External gear pumps are a three-piece construction consisting of two end covers and a central housing held together by bolts. The housing accommodates two identical gears in constant mesh with each other, which are free to rotate. The end covers also act as bearings for gear shaft journals. The driving gear is an extension of the drive shaft. As the teeth un-mesh a partial vacuum is created enabling the flow of oil from the reservoir to the suction port from where successive pockets between the teeth convey this oil to the discharge port from where engaging teeth force the oil out. The material properties and manufacturing accuracies of the housing is critical to the volumetric efficiency and durability of this pump. Pumps made of aluminum alloy housing are rated for higher pressures as compared to cast iron housings since the body to gear geometry can be obtained during the running in test period thus obtaining zero clearance between the gear tips and housing.

This ensures a perfect gear tip seal under actual running conditions. Since aluminum alloys are not as hard and as wear resistant as cast iron, the durability of pumps made of aluminum alloy housing is less than that of pumps made out of cast iron housing.

Although different manufacturers rate their products as good for pressures up to 250 kgf/sq-cm, gear pumps are not recommended for pressures beyond 150–175 kgf/sq-cm especially, if the duty cycle is continuous over a period of 8 hours. Pumps with flow ratings, up to 200 lpm are available.

Spur gears are the most commonly used gears in gear pumps. Although helical gears can withstand greater tooth load, run more smoothly and at higher speeds, they can transfer less fluid per tooth as compared to the spur gear. The displacement volume of a gear pump depends on the depth of the tooth and the width of the gear and is given by,

$$Q_d = 2\pi D_p b m$$

where, Q_d is theoretical pump displacement in cm^3 per revolution, D_p is the pitch circle diameter of the gear in centimeters, b is the gear width in centimeters, and m is the gear module in centimetres.

Outlet pressure against the teeth causes heavy side loads on journal bearings. Therefore, direct drives using flexible couplings is recommended for gear pumps since it is not feasible to impose additional side loads on the bimetal bearings such as those caused by belt or chain drives. Pressure balanced pumps rated for much higher pressures have been developed but they offset the price advantage that ordinary gear pumps enjoy.

Gear pumps are the "use and throw" kind of equipments since neither the gears nor the pump housing can be salvaged. Since the gears are matched to their housings in assembly, interchangeability is a problem and the only spare part that can be used is perhaps, the seal kit. Wear out of bimetal journal bearings and the gear housing individually and collectively contribute to the deterioration in performance resulting basically in its incapability to build up rated pressures. Therefore effective contamination control both at the entry point to the pump and at the return lines becomes critical. Since the differential thermal expansion between the aluminum pump housing and gear teeth can cause increased clearance and consequent reduction in performance levels, control of oil temperature too, becomes critical for this pump due to the cumulative effect of increased clearance and reduced oil viscosities. High performance gear pumps are available but they cost more and hence their price advantage is lost.

The running noise of gear pumps is in the region of 95–100 dB, perhaps the highest among positive displacement pumps except for axial piston pumps. They are available only as fixed delivery pumps and the only way the delivery can be varied is by varying the RPM of the prime mover driving the pump. This is possible if the prime mover is an IC engine. These pumps therefore are widely used as very economical alternatives to variable delivery pumps in mobile applications where the prime mover is usually a diesel engine. They are also used in industrial hydraulic systems in low and medium pressure applications since they cost less than other pumps.

Internal gear pumps (Fig. 2.3) have two dissimilar gears in mesh, one inside the other. Of the two, the driving gear is smaller and is an external spur gear, where as the driven gear is an internal spur gear. A crescent shaped seal separates the inner and the outer gears. Unlike external gear

Fig. 2.3. *Internal gear pump.*

pumps, the two gears in the internal gear pump rotate in the same direction but at different speeds, the inner rotating faster than the outer. Their running noise is much less than external gear pumps.

The mechanism of pumping action is similar to external gear pumps. Internal gear pumps make much less noise than any other type of pumps. Their volumetric efficiencies are higher and therefore they are rated for higher pressures compared to an external gear pumps.

2.4 VANE PUMPS

The modern fixed displacement vane pump consists of a rotor splined to a drive shaft. The rotor has equally spaced slots milled on its periphery. Rectangular pieces of hardened and ground alloy steel strips called vanes whose thickness is closely matched to the slots in the rotor freely slide inside the slots. The whole assembly is located central to a ring called the 'cam ring' which has an elliptical bore. The cam ring is press fitted to a housing with side plates on either side. The housing is held in position by two end covers which also act as housing for shaft bearings. The mounting bracket is integral with the front-end cover.

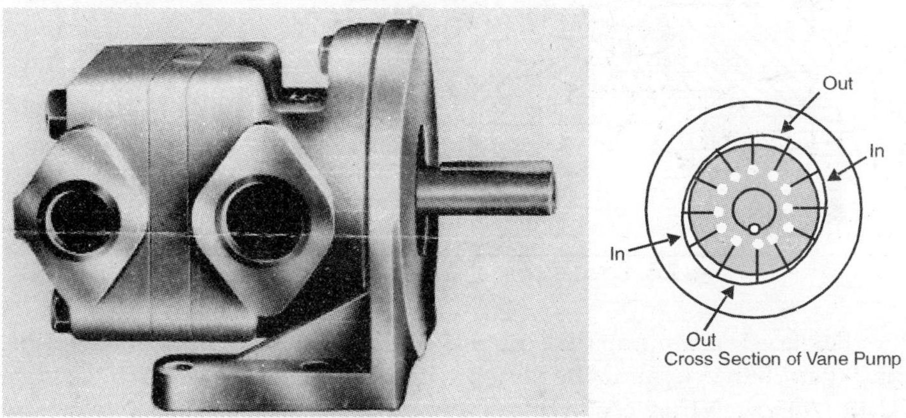

Fig. 2.4. *Vane pump.*

As the rotor rotates inside the cam ring the vanes are forced outwards due to the centrifugal force and are held against the surface of the cam ring which due to its eccentricity forces the vanes in and out of the slots. During rotation, as the chamber volume increases the vanes extend and sweep the oil entering the chamber due to the vacuum created.

This swept volume is forced out of two pressure chambers, to the outlet port. Since the two pressure chambers are exactly 180 degrees apart side loads on the rotor shaft cancel each other out and the shaft is in balance. This pump has therefore come to be known as the *fixed delivery balanced vane pump* and is an improvement over its earlier version, the unbalanced vane pump.

The theoretical displacement of a fixed delivery vane pump is given by,

$$q_t = 2\,\pi\,L\,b\,D \text{ cm}^3 \text{ per revolution.}$$

where, L is the vane stroke or throw in cm, b is the breadth of the vane in cm, and D is the diameter of the rotor in cm.

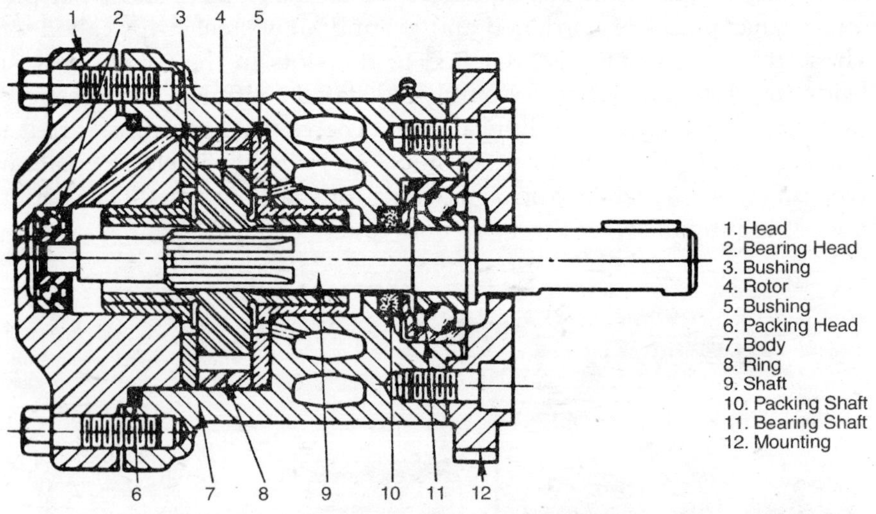

Fig. 2.5. *Cross section of fixed displacement vane pump.*

Balanced vane pumps like the gear pumps are fixed displacement pumps and their delivery cannot be varied without varying the RPM of the pump shaft. Although they cost a little more initially, compared to a gear pump their service life is much longer than a gear pump because, they can be re-conditioned by replacing a few worn out parts. Vane pumps maintain their efficiency for a long time because, wear of the vane tips is automatically compensated by vanes, which slide further in their slots to maintain contact with the housing. Their noise level is around 75–80 dB, perhaps the lowest compared to other pumps.

They are available with a discharge rating up to 150 LPM and are good for pressures up to 200 kgf/sq-cm.

The two side plates, also called pressure plate and wear plate together with cam ring, the rotor, vanes, mounting screws and locating dowels, form one sub assembly and the unit is called the cartridge, which is available as a replacement item. The pump flow rating can be increased or decreased simply by replacing this cartridge. Vane pumps are less sensitive to contaminants than a piston pump. They require a minimum of 600/700 RPM for the vanes to engage. They operate efficiently up to 2000 RPM. Unlike gear pumps their inlet to outlet port relationship is not fixed once assembled. Their relationship can be changed even during installation to suit ones requirement simply by removing the tie bolts and rotating the covers. This flexibility eliminates problems during installation and piping. Also they can be assembled for clockwise or anticlockwise rotation. Although co-axial drive with flexible coupling is recommended for these pumps, they can take side loads imposed by a belt or a chain drive better than the Gear pump and so indirect drives may be considered if allowance can be made for the additional load imposed on pump shaft bearings.

2.5 PISTON PUMPS

For pressures beyond 200 kgf/sq-cm piston pumps today are the only choice and their volumetric efficiency is the highest compared to other pumps.

The hand pump, about which we have already discussed, is a piston pump in its simplest form. It is a single or double piston arrangement reciprocating in a closely matched cavity actuated by hand force. These pumps can meet any pressure requirements but can hardly meet industrial flow requirements.

Power driven piston pumps on the other hand have three or more pistons reciprocating in their cavities at high speeds actuated by a wobble plate or a cam/eccentric depending upon whether it is an axial or radial piston pump. In axial piston pumps, the pistons reciprocate parallel to the axis of the drive shaft and in radial piston pumps the pistons reciprocate perpendicular to the axis of the drive shaft. Some of the Piston pumps built for industrial applications such as the swash plate piston pumps and radial piston pumps of the oilimmersed type are competitive price wise, with high performance gear or vane pumps and therefore can be considered as an alternative as they offer 'higher pressure' advantage.

In axial piston pumps an angled cam also called a wobble plate is integral with the drive shaft. Rotation of the wobble plate causes forward linear motion of the pistons in contact, the return motion being achieved by compression springs. The fluid is drawn in while the piston moves

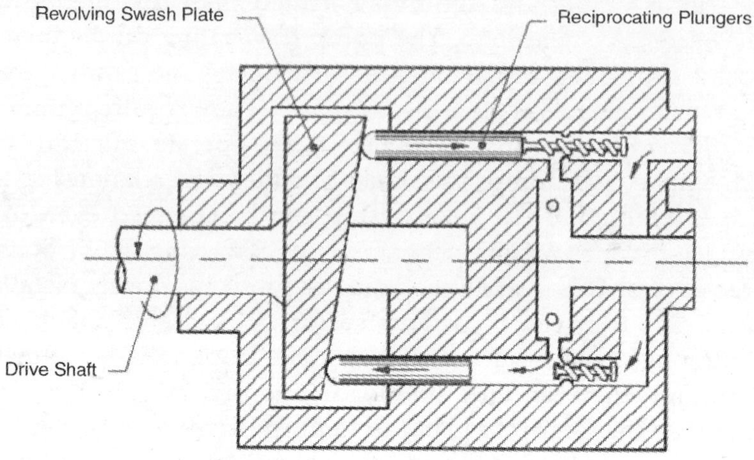

Fig. 2.7. *Axial piston pump cross section.*

towards the thin section of the plate and is expelled while approaching the thick end. These pumps use individual check valves to regulate the flow in and out of the pistons. These pumps are rugged, durable and can rotate in either direction. They resist cavitations better and they are usually the choice in applications involving high operating pressures and large deliveries. They are available in pressure ratings up to 500 kgf/sq-cm. The type shown in Fig. 2.7 is that of an axial piston pump with rotating swash plate and a stationery cylinder block in which the plungers reciprocate. A variation to this design is the inline piston pump in which the cylinder block is driven by the drive shaft and the pistons reciprocate against a stationary wobble plate. These pumps regulate the inlet and out let flows with the help of a valve plate having two kidney shaped ports. They are easy to maintain, permit flow reversibility and are good for moderate flows and pressures within the limits, 200–220 kgf/sq-cm.

Radial piston pumps may be either of the check valve or pintle-valve type. In check valve type a rotating cam that is integral with the drive shaft imparts reciprocating motion to the pistons. Check valves control fluid in and out of the pistons. In pintle valve pumps the cylinder block and pistons together are free to rotate concentric to the center of a pintle. The rotation of the cylinder block and the pistons is eccentric to a stationary reaction ring. As the block rotates, the pistons reciprocate in their bores. While the inward motion is due to the offset of the rotating block with respect to the center of the reaction ring, the out ward motion is due to the combined action of the charging pressure and the centrifugal force. Porting in the pintle permit the fluid to be drawn in as the pistons move out ward and expelled, as the pistons move inwards. Cam actuated radial piston pump has the pressure and flow rating, up to 700 kgf/sq-cm and up to 500 LPM. Pintle valve types are good for pressures, up to 200–220 kgf/sq-cm.

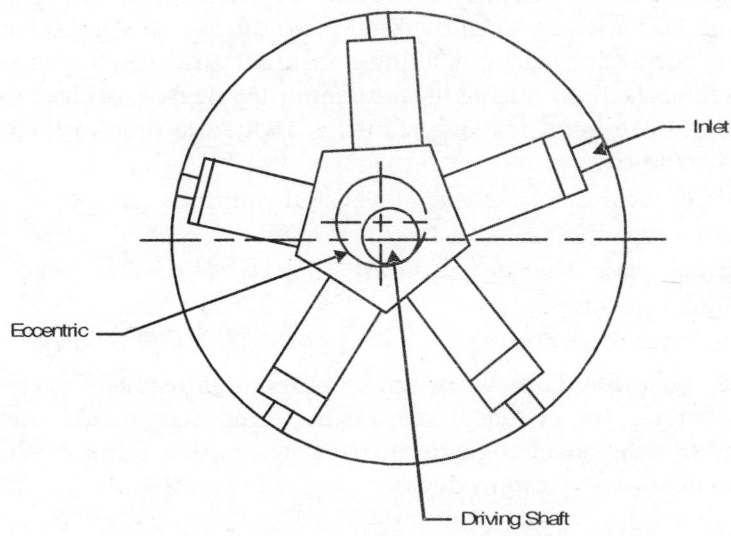

Fig. 2.8. *Mechanism of action of radial piston pump.*

Fig. 2.9. *Radial piston pump.*

Another variation of a piston pump is the bent axis piston pump, so called because the axis of the pumping chamber consisting of the cylinder block, the pistons and the valve plate is inclined at an angle to the drive shaft. The drive shaft transmits motion to the cylinder block through a ball and socket joint and the reciprocating pistons are hinged to the drive shaft flange, again through a ball and socket joint. Because the plane of rotation of the pistons is at an angle to the valve plate, the distance between the pistons and the face of the valve plate keeps changing during rotation. The pistons move away from the vale plate during 180 degrees of shaft revolution permitting vacuum filling and move towards the valve plate thus expelling the fluid during the remaining 180 degrees of shaft rotation. These pumps are good for large flows at moderate pressures, between 200–220 kgf/sq-cm.

The theoretical displacement of a piston pump in cm^3 per revolution is given by

1. Swash plate axial piston pump $Q_t = 0.7854\ d^2\ Z\ D_c \tan\alpha$
2. Radial piston pump $Q_t = 0.7854\ d^2\ e\ Z$
3. Bent axis piston pump $Q_t = 0.7854\ d^2\ Z\ D_c \sin\alpha$

where d is the piston diameter in cm, Z= number of pistons, D_c is the pitch circle diameter of the pistons in cm, e is the eccentricity or the cam throw in cm, and α is the offset angle in case of a swash plate pump or the angle of tilt in a bent axis pump in degrees.

2.6 TANDEM PUMPS

These are two pumps with a common shaft connected in series with two different pumping chambers contained in the same housing. They will have two discharge outlets and a common inlet.

They are available as gear, vane or piston units. These or compact and ideal pumps for 'high-low' circuits designed primarily to conserve input HP. The delivery from the two pumps can be combined to achieve the required approach or return speed of the actuator. When the demand for full flow ceases, that is, when the actuator stops movement, the larger of the two pumps can be unloaded to tank at low pressure.

Fig. 2.10. *Tandem pump.*

The low delivery pump can now meet the systems pressure requirement thereby reducing the input power required. Some times two different pumps, one of which has the drive shaft at either end is be mechanically coupled to the other to achieve the same purpose.

2.7 SCREW PUMPS

These are ideal pumps for low pressure and very high flow requirements. They can maintain a continuous pressure between 30–60 kgf/sq-cm while delivering fluid up to 7000 LPM. Their volumetric efficiency at this pressure is between 85–90 percent. These pumps are quite running as there is no metal-to-metal contact, are more tolerant to contamination and are highly reliable. They are basically axial flow pumps with non-pulsating delivery. They can be of single screw, double screw or three screw arrangement.

A typical three-screw pump consists of a central drive shaft with an integral screw, which is in mesh with two idler rotors. The entering fluid is trapped between the rotors as they rotate and is conveyed to the discharge end and finally expelled. Factors influencing the choice of a constant delivery pump.

2.7.1 Factors Influencing the Choice of a Constant Delivery Pump

External gear pumps cost less than other pumps but their volumetric efficiency is the lowest compared to other pumps, between 65–85. This means for the same discharge and pressure the input power required would be higher as compared to other pumps whose efficiencies are higher. Also, their running noise is high compared to other pumps. Pumps currently manufactured in India are no good for pressures beyond 120–150 kgf/sq-cm if the duty cycle is continuous. Some of the best gear pumps are made in USA and Germany. These pumps are good for pressures up to 200 kgf/sq-cm, have a higher volumetric efficiency and higher life expectancy but costs more. So for applications requiring moderate pressures and low duty cycles gear pumps are the right choice mainly because they cost less. They are normally recommended for flow requirements between 5–150 liters per minute.

Internal gear pumps have a higher pressure and flow ratings compared to external gear pumps. They are also more efficient but they cost more. Also they are not currently manufactured in India and so needs to be imported.

For low operating pressure (30–60 kgf/sq-cm) and huge flow requirements (500–100 LPM) screw pumps are the ideal choice. They are very reliable and quiet running and are easy to maintain. Vane pumps especially of the balanced vane type are more reliable, have a long service life and are quiet running compared to the gear pumps.

Vane pumps made in India are rated for pressures between 140–175 kgf/sq-cm under continuous duty cycles and are available in flow ratings between 5–200 LPM. Their volumetric efficiency is between 85–90%. They cost 50% more than the gear pumps. Their higher initial investment can however be justified by their higher-pressure ratings, longer service life.

For very high pressure applications between 200–400 kgf/sq-cm and beyond there is no alternative to piston pumps. They are highly reliable, have very high life expectancy, their volumetric efficiency is the highest (90–95%) and they are available in discharge ratings from 5–500 LPM. Among pumps of the same class, higher the flow rating higher will be the initial cost of the pump. Depending upon their pressure rating and flow rating they cost anywhere between 2–20 times more than a gear pump of the same flow rating.

If the flow requirement is moderate but high-pressure requirement exists then radial piston pumps may be the ideal choice. If the flow requirement is also large than the swash plate axial piston pump may be the best choice both from the point of view of availability and cost competitiveness.

2.8 VARIABLE-DELIVERY PUMPS

Pumps hitherto discussed were constant delivery pumps. This means, that they deliver a fixed quantity of fluid regardless of actuator requirements during its entire movement. We have discussed earlier, that the two main characteristics of the pump used in oil hydraulic systems are, pressure and delivery. While pressure capability determines the out put forces or torque developed by the actuator, the pump delivery, controls the linear or rotational speed of the actuator depending upon whether the actuator is a hydraulic cylinder or a hydraulic motor.

The product of pressure p and the delivery Q is directly proportional to the power in put, required to drive the pump. The philosophy of every hydraulic system designer is to ensure that the input power is optimum and wastage of energy is minimum. By appropriate circuit designs it is possible to ensure an optimum input power based on the duty cycle of the machine. This will be discussed in depth separately with suitable examples as a part of application engineering. What is discussed here is not optimum in put power but optimum consumption of energy with minimum wastage.

In a hydraulic system pressure developed by the pump is proportional to load resistance and hence the work done or energy consumed at any point is a product of load resistance offered at that point (Δp) and the pump flow Q, which is constant at all points. This has obviously lead designers to the thought that if flow requirement could also be similarly conditioned to meet the duty cycle requirement, then an ideal situation would emerge resulting in appreciable savings in, what would otherwise constitute, a wasted energy. Also large quantities of fluid at high pressure blowing over the relief valves during that part of the cycle when the actuator movement stops due to maximum load resistance would contribute to heating up of oil and consequent damage to the hydraulic system. The development of *variable delivery pumps* was precisely to over come this problem. Load resistance conditions the delivery from this pump. Higher the load resistance lower will be the delivery to the extent that the delivery will be almost zero at maximum system pressure. This means the energy consumed is minimum even while maintaining maximum system pressure requirements. Since the pump flow is need based, dumping of large quantities of oil at high pressure once during every cycle is avoided.

The higher initial cost of the basic variable delivery pump can be justified on the basis of improved efficiency of the hydraulic system and long-term gains in energy costs. The input power required is still based on maximum pressure and maximum flow required by the circuit. This is because the 'peak power' requirements must first be met before the pump begins to regulate the flow. Since this time interval is short in some cases it is possible to reduce the HP of the electric motor by 20–30 percent by coupling a suitable flywheel to the motor shaft so that the accumulated energy in the flywheel can assist the electric motor to drive the pump during this period.

Vane pumps of the unbalanced type, axial piston pumps with rotating cylinder block and stationary swash plate, bent axis piston pumps and radial piston pumps of pintle valve type are the types considered suitable for incorporating variable delivery features.

In variable displacement unbalanced vane pumps the displacement control mechanism employed, moves the cam ring to change the eccentricity between the ring and the rotor thus changing the volume of the pumping chamber which in turn effects the displacement per revolution. When the pressure is high enough to over come the compensator setting (spring force), which is adjustable, the cam ring shifts to decrease the eccentricity.

In piston pumps with rotating cylinder block and stationery swash plate the angle of the swash plate installed in a movable yoke pivoted on pintles is varied. This alters the stroke of the plungers and in turn the pump flow rate. The positioning of the yoke at a pre-determined angle to the axis within the angle of tilt permitted by the design can be accomplished either by turning a hand wheel or by shifting a lever manually. Hydro-mechanical, Electro-hydraulic or servo control mechanisms have also been developed and therefore it is now possible to actuate these displacement control mechanisms from a remote location. These pumps have come to be designated as *non-pressure compensated pumps* because they are not the self-regulating type and require auxiliary displacement control mechanisms to regulate the flow. In non-pressure compensated pumps it is possible to reverse the discharge connection by reversing the direction of drive shaft rotation.

Pumps using Hydro-mechanical or Electro hydraulic or servo controls as displacement control mechanisms require an auxiliary pump to actuate the valves and these are available as integral units mounted on the pump housing and driven by the internal 1 : 1 drive by the main pump shaft. In *electronic controls* adjustment of the position of cam ring in a vane pump or the yoke angle in an axial piston pump is achieved by the use of proportional solenoids or proportional pressure controllers.

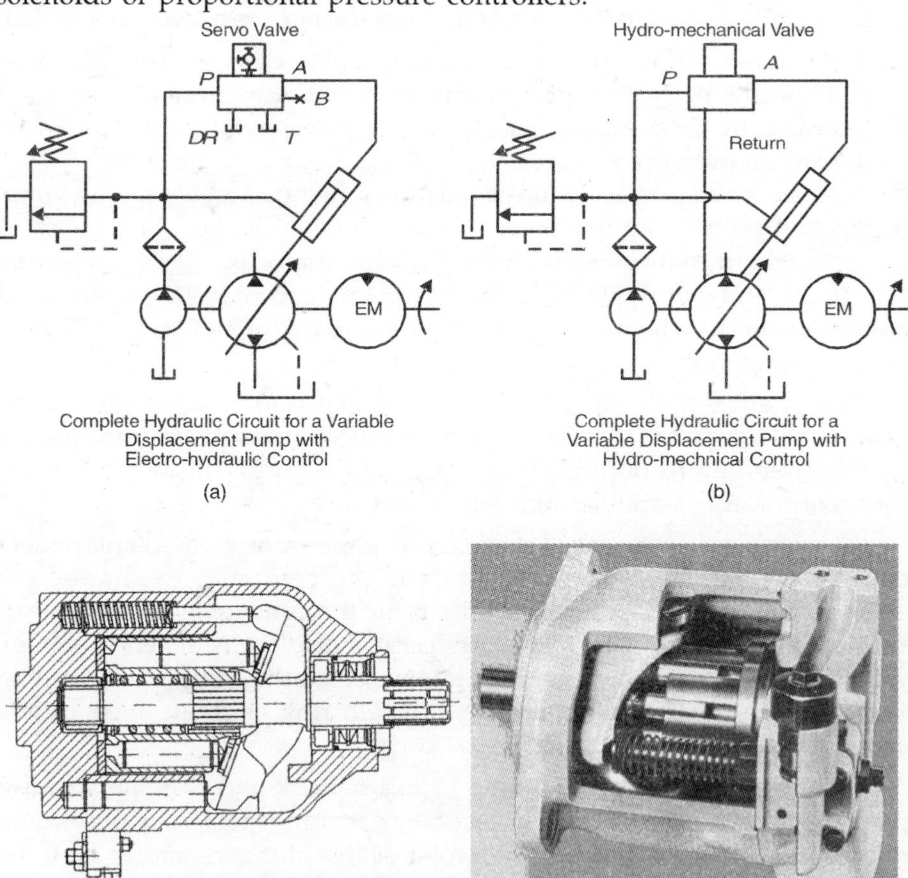

Fig. 2.11. *A variable delivery axial piston pump.*

These convert input signals such as amplified current or a pulse width modulated signal into a proportional force sufficient to displace the cam ring or the yoke from its existing position thus varying the pump displacement. Non-pressure compensated pump controls can be termed 'direct displacement control systems' and are normally preferred when the pump drives a single load. Variable delivery pumps designed for direct displacement control are energy efficient, and can pump in either direction permitting reverse flow. The total input HP required is the HP required for the main pump plus the HP required for the auxiliary pump based on their individual pressure and flow rating. Since direct displacement pumps are invariably used in closed circuit drive they require a booster pump to make up 'leakage losses'.

Pressure compensated pumps are also called self-regulating pumps because they do not need an auxiliary displacement control mechanism. Instead a pressure-compensating device, which is integral with the pump senses the system pressure and de-strokes the pump at the 'cut-off ' pressure, which can be pre-set. To derive maximum benefit, these pumps have to be used in closed center valves.

There are many variations to pressure compensator controls but the most familiar are the 'flat cut-off' compensator and the flat cut-off compensator with limit stop. In a flat cut-off compensator control, below the threshold pressure the pump flow is constant and above the threshold the pump flow falls rapidly in a single steep step to zero. In a flat cut-off compensator with a limit stop a mechanical stop limits minimum pump displacement to some small but positive value. Because displacement cannot go to zero a relief valve is required to control maximum pressure.

An improvement over the simple pressure compensation control is the *load sensing control*, which instead of sensing the absolute pressure senses the pressure drop across a variable orifice and the load sensing compensator spool thus sensing both the outlet pressure and the load pressure. The compensator varies the pump stroke so as to maintain a constant pressure differential. Unlike in simple pressure compensator controls, in load sensing controls the load velocity remains constant even if the load increases. The wasted energy is appreciably less in variable delivery pumps equipped with a load sensing compensator control as compared to pumps equipped with simple pressure compensator control.

In mobile hydraulic systems reduced speed at higher loads is preferred such as in lift trucks, loaders, cranes etc. For such applications variable delivery pumps equipped with *'torque limiting' controls* provide an effective solution. This control system allows full flow velocity at moderate loads even while retaining the ability to move high loads at reduced velocities. This is the only system that permits selection of a smaller prime mover and is therefore called the *horsepower limiting plus energy saving control.*

Other less popular hydro-mechanical control systems in use are, speed sensing, torque summation, and torque select controls not commonly used in industrial hydraulic applications.

The degree of sophistication that can be introduced to these controls is, limited only by the expected levels of linearity of flow, pressure and repeatability. A closed loop feed back control using a pressure transducer for pressure monitoring and an LVDT for flow monitoring may be required to impart a high degree of accuracy to these parameters.

A major advantage of variable delivery pump controls especially the closed loop electronic feed back control lies in appreciable reduction, about 40%–50% in energy and operating costs. Further, they simplify the hydraulic circuit by eliminating the need for flow control and pressure control valves and two pump systems. Troubleshooting becomes easy because power elements and controls are isolated. Because it is possible to exercise precise control over acceleration and deceleration cycle times can be decreased and the need for a deceleration valve is eliminated. The actuators can be sized to meet the maximum load requirements instead of maximum speed requirements.

2.8.1 Constant vs Variable Delivery Pumps: Influencing Factors

If the input HP is less than 20 or if the duty cycle is on-off type enabling unloading of the pump discharge to tank during the valve neutral position or oil temperature raise and energy costs is not a severe problem, constant delivery pumps are considered the appropriate choice.

If the input HP is above 20, if the pump has to serve more than one load at different pressures and at high frequencies and if small precise motions under high loads and rapid motions under varying loads are required a variable displacement pump can be considered. The degree of sophistication of controls for variable delivery pumps should however be dictated by system needs and economic considerations. Large systems with input power more than 100 HP would absorb the higher costs of electronic controls better than smaller systems that may have to relay on more elementary controls such as displacement controls, pressure compensator controls, etc.

3

POWER INPUT DEVICES: ACCUMULATORS AND INTENSIFIERS

3.1 ACCUMULATORS

In oil hydraulic systems the function of an accumulator can be compared to that of the flywheel used in mechanical presses. Its function is also analogous to the use of a car battery for providing high current for short period required during starting.

In hydraulic presses having duty cycles similar to mechanical presses a flywheel can be coupled to the pump and the motor shaft to reduce the input HP of the electric motor. During peak power requirements the flywheel energy would drive the pump. The electric motor in this case may be sized to overcome the pressure drop and the pressure required to return the actuator to its original position.

Because flywheels are a source of kinetic energy, their versatility and adaptability to various applications is seriously limited. Their energy cannot be tapped at 'will', whenever and wherever required. But accumulators used in oil hydraulic systems are a 'potential energy' source. Because of this, their use is limited only to the ingenuity of the hydraulic system designer. They acquire this energy from the pump during the lean or idling periods of the system, for release during peak flow or sudden energy demands of the system. Thus they can supplement pump flow enabling system designers to select a smaller pump or they can perform the pressure holding function relieving the pump for other purposes or enable the electric motor to be disconnected thus conserving energy.

Their basic function therefore is to conserve energy, either as volume compensators or as pressure compensators. As pressure or volume compensators they are used to perform any of these tasks.

An accumulator can act as an auxiliary power source where duty cycle is intermittent with a small pump acting as the primary source.

Cavitations of piston pumps can be prevented, by installing an accumulator close to the pump suction line. The accumulator pressurizes the inlet, reduces the effective length of the inlet line and can absorb excess kinetic energy.

Dual pressure circuits that permit fast motion at low-pressure during the initial phase and slow motion at high pressure towards the end of the stroke can effectively use accumulators.

Slow motion at high pressure can be supplied by a high-pressure low-volume pump, which can also be used to re-charge the accumulator during the 'hold-on' portion of the cycle. When rapid extension is required fluid is expelled from the accumulator to assist the pump in extending the piston fast.

An accumulator can act as a source of 'make-up' fluid where the pressure must be held for long periods with the pump cut-off or the flow diverted to meet system requirement elsewhere. The accumulator under these circumstances will make up the inevitable leakage losses and holds the pressure steady.

Pressure pulsations that can originate especially in a piston pump can be reduced, by locating an accumulator close to the pump outlet. These pulsations are the primary source of noise and vibration in a fluid power system.

If an accumulator is considered early in the design process, it can help reduce electric motor and pump sizes and thus help save energy. By using an accumulator to supplement pump delivery a smaller pump and electric motor can be selected thus reducing input HP and long term energy costs.

An accumulator can perform the function of a counter weight in machine tools, hoists etc. In this application they save 'weight' and considerable space.

In applications involving transfer of large amount of fluid from one location to another, especially over a gradient sudden closure of valves can produce 'shock waves' which results in pipe burst, joint separation, valve leakage etc. This problem can be effectively tackled by the use of an accumulator, which acts as a hydraulic 'cushion' or as shock absorber.

They are used as a 'standby' to provide emergency source of power during power failures or during hydraulic system failures so that reserve energy is available to complete the cycle and return the actuators to their normal positions. They are also used to accommodate volume increase due to thermal expansion of fluid in a closed circuit hydraulic system.

They can perform the function of mechanical extension or compression springs more effectively and where the higher costs justify the function such as in deep drawing presses where they are used as alternatives.

There are two types of accumulators. The mechanical and the Hydro-pneumatic.

Mechanical accumulators are the earliest versions of accumulators. They were either the weight loaded type or the spring loaded type. They are now, merely of academic interest having lost their ground to the more efficient and compact hydro-pneumatic variety.

In weight-loaded accumulators pump delivery when admitted to the chamber underneath the piston overcomes the resistance offered by weights causing the piston to rise, thus creating a potential energy source. Weight loaded accumulators are the only type that can provide a constant pressure source regardless of change in the volume of the fluid as the weights descend during energy release. Weight loaded accumulators are heavy and bulky. A 20 liters capacity will weigh about 3000 kg as compared to a weight of 75 kg of a modern Hydro-pneumatic accumulator. This is a serious disadvantage. In applications requiring a constant pressure source or very large volumes of fluid they may still be considered as an alternative.

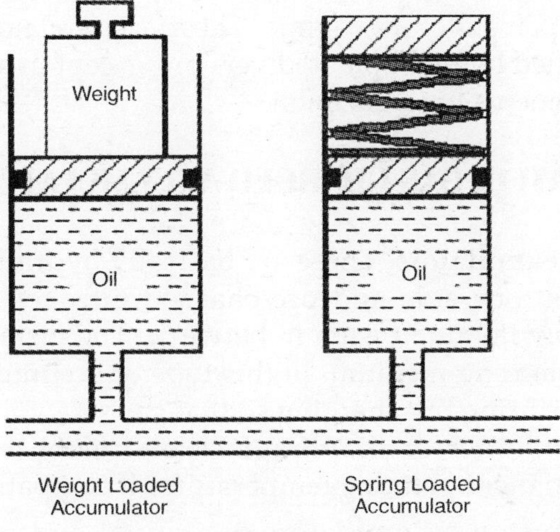

Fig. 3.1

In spring loaded accumulators the energy is stored by compressing a helical spring located behind the accumulator piston. The reaction force on the fluid is equal to the product of spring constant and the total deflection suffered by the spring. These accumulators are not economical for large flow or pressure requirements since for large flows, the size of spring required becomes unwieldy.

In hydro pneumatic accumulators energy is imparted to the hydraulic fluid by compressing a 'pre-charged' gas instead of compressing a spring or raising a load. This gas is always nitrogen because nitrogen is inert and therefore there is no fear of combustion even if the gas gets heated. There are two types of hydro-pneumatic accumulators. The *separated* and *the non-separated types*.

In the non-separated type the hydraulic oil and the gas are practically in physical contact as there is no intermediate chamber separating them. In the separated type either a floating piston or a diaphragm or a rubber bladder separates the hydraulic fluid and the gas and accordingly they are called 'piston accumulators', 'bladder accumulators' and diaphragm accumulators.

The non-separated type of hydro-pneumatic accumulators are more simple, costs less and can store a larger volume of fluid than the separated type. They respond faster, to the system demand. It is important in this type to ensure that not more than sixty five percent of oil is consumed at any time to prevent accidental discharge of gas in to the system. These have to be mounted vertically to keep nitrogen, which is lighter, always above the hydraulic oil. The main drawback of this type of accumulators is that the gas may get mixed with oil in course of time. Prior to pre-charging these accumulators, it is important to fill up the chamber with oil up to 30–40 percent of its volume. These accumulators have now been largely superseded by the separated versions except in applications where volume requirements are very high.

3.2 SEPARATED, GAS CHARGED ACCUMULATORS

1. Piston accumulators. These are basically hydraulic cylinders with hydraulic oil on one side and pre-charged nitrogen gas on the other side with a freely floating piston in between. The piston in this case acts as the fluid-separating medium. In this type of accumulators the oil side must be empty during pre charging so that gas side volume is maximum. Compared to bladder or diaphragm type, piston accumulators can accommodate higher operating temperatures if compatible seals are used.

Failure of piston accumulators is gradual and is more often due to 'wear out' of piston seals. Worn out piston seals permit leakage of high-pressure hydraulic oil to the gas side with each stroke thus increasing the pre-charge pressure. As the pre-charge pressure increases the storage and exhaust capability decreases. If we fail to set right this deterioration by replacing seals, the accumulator will fail totally when in course of time the pre-charge pressure becomes equal to the maximum system pressure. Piston accumulators are not recommended for pulsation damping because of piston inertia and sidewall friction between the cylinder and the piston.

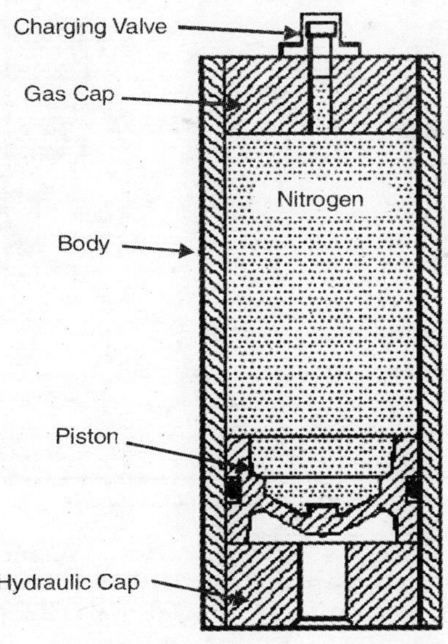

Fig. 3.2. *Piston accumulator.*

2. Bladder/diaphragm accumulators. Diaphragm accumulators are usually spherical in shape with a synthetic rubber diaphragm separating the oil from the gas side. Their capacity is usually limited to 2-3 litres. Bladder accumulators have a cylindrical shell instead and has a rubber bladder in which the gas is contained. Hydraulic oil fills the space around it.

Failure of bladder/diaphragm accumulators is total and without warning and occurs due to bladder rupture. This feature is considered a blessing in disguise in high volume production machines where product quality depends on system pressure. In such applications, sudden failure becomes obvious forthwith and steps can be initiated to solve the problem thus eliminating scrap. Vertical mounting with hydraulic port down is preferred for both piston and bladder accumulators. Piston models can be mounted horizontally if need arises.

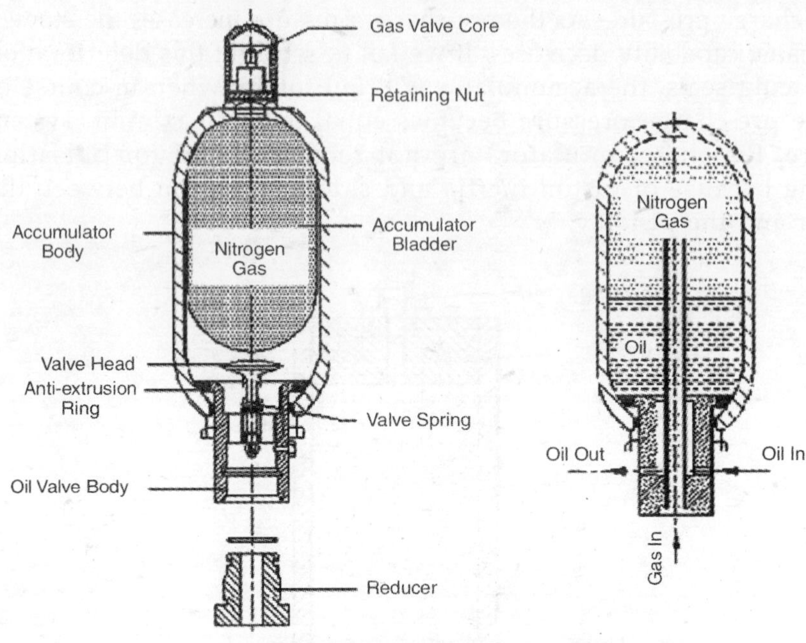

Fig. 3.3. *Non-separated accumulator.*

Horizontal mounting of bladder accumulators can cause uneven bladder wear but is permitted when it is being used as a pulsation damper or when it is performing a pressure holding function. It is important in case of bladder accumulators to maintain the designed pre-charge pressure before admitting high-pressure hydraulic oil in order to prevent crushing of the bladder or extrusion of the bladder in to gas valves. This does not lead to such severe problems in case of piston accumulators. Maintaining the pre-charge pressure as specified by the manufacturer is very important for efficient functioning of the system and for prolonging the life of the bladder.

In a typical energy storage application a bladder accumulator is pre-charged to 80 percent of *minimum* system pressure. Piston accumulators are pre-charged to a pressure, which is about 7 kgf/sq-cm less than *minimum* system pressure. Accumulator charging begins when hydraulic fluid is admitted through the oil port after the gas side has been pre-charged. The oil now compresses the gas and stores energy. Since maximum operating system pressure and minimum operating system pressure have a direct bearing on the gas pre-charge pressure they should not be varied from their pre-determined values without a corresponding variation in gas pre-charge pressures.

Bladder accumulators are designed as a pressure vessel and should be treated as such. The shell is made out of high strength steel if it is meant for mineral oil. In case of corrosive fluids it could be stainless steel. The flexible bladder is manufactured out of a special process and is a molded rubber single piece without seams. The gas valve is sometimes vulcanized to the bladder or is fitted in a manner, which allows it to be connected and disconnected easily and safely. Not being an integral part of the bladder, it can be reused thus reducing maintenance costs. The puppet valve on the oil side is designed to close automatically during sudden dropping of the system pressure preventing extrusion of bladder material and consequent damage. Standard accumulators are designed to operate safely between temperatures –10ºC to +80ºC.

In an accumulator, at any point of time, we are either compressing a pre-charged gas or allowing it to expand. This compression or expansion brings about a status change in the gas, which is governed by the perfect gas equation,

$$PV = WRT \tag{1}$$

where P is the absolute pressure in kgf/sq-cm, V is the gas volume in cm^3, W is the mass in kg, R is the universal gas constant which for gaseous nitrogen is 30.26 kg-m/kg ºK. For the particular gas and the accumulator, the value of WR is constant and Eq. (1) can written as, PV/T = constant or,

$$P_0V_0 / T_0 = P_1V_1 / T_1 \tag{2}$$

When the change takes place over a long period of time the temperature of the gas remains constant and such a change is called *isothermal*, resulting in the equation,

$$P_0V_0 = P_1V_1 = P_2V_2 \tag{3}$$

When the change occurs instantaneously there is no time for heat transfer from the work to the environment and such a change is called isentropic or reversible adiabatic and is given by,

$$P_0 V_0^n = P_1 V_1^n = P_2 V_2^n \tag{4}$$

All changes between isothermal and isentropic are called polytropic.

Pressure-volume diagram shown in Fig. 3.4 will help us to understand how the volume variation as a function of pressure is depending on the value of the polytropic exponent n which for nitrogen is contained within the limits $1 \le n \le 1.4$. The value of n is taken to be equal to 1 if the compression and expansion process takes place under isothermal process For adiabatic conditions the value of n is taken equal to 1.4. Polytropic values of n can be obtained from the graph of n versus time t in Fig. 3.5.

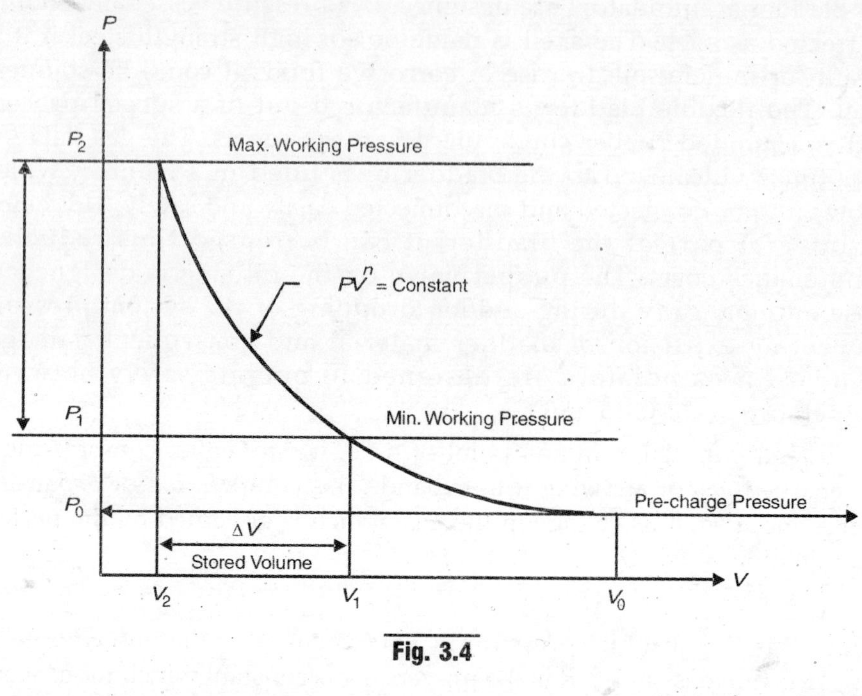

Fig. 3.4

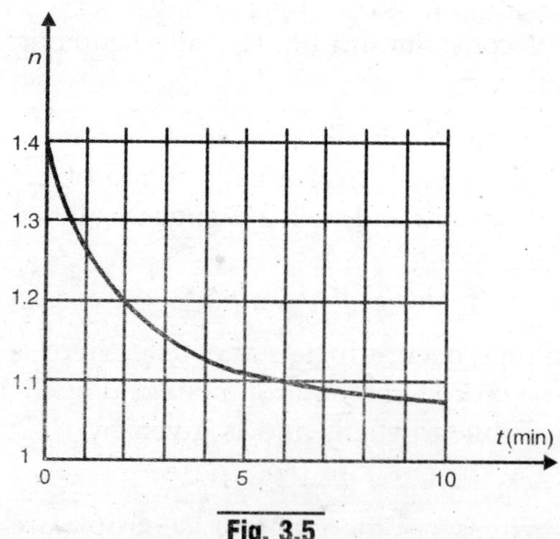

Fig. 3.5

Isothermal conditions can be considered to exist if the accumulator is used as a volume compensator, leakage compensator and pressure compensator or as a lubrication compensator. In all other cases such as, energy accumulation, pulsation damping, emergency power source, dynamic pressure compensator, shock absorber, hydraulic spring etc., expansion and compression process may be considered to take place under 'adiabatic' conditions. Generally, adiabatic condition is considered to exist if the compression or expansion period is less than 3 minutes.

3.3 ACCUMULATOR SELECTION

After ascertaining the type of accumulator that is appropriate for the purpose envisaged, what remains is determining the volume of the accumulator in liters that can effectively supply the required energy in the form of 'high pressure fluid'. Accumulators are manufactured to a variety of pressure ratings and the one chosen should be rated for a pressure more than the maximum system working pressure P_2.

However the values of the following basic parameters should be established before proceeding further.

Working pressures P_1 and P_2. The value of P_2 is found from the ratio, $P_2/P_0 \leq 4$. The maximum gas pre-charge pressure is found from the relationship $P_0 \leq 0.9\ P_1$ or $P_0 \geq 0.25\ P_2$. The gas pre-charge pressure must be as close as possible to the minimum working pressure P_1 to obtain maximum storage. Special values for P_0 are used in pulsation damping and shock absorber applications ($P_0 = 0.8\ P_1$). Other parameters to be determined are the, volume of fluid ΔV that needs to be stored ($\Delta V_{max.} = 0.75 V_0$), the maximum required flow rate and the operating temperature.

3.4 SIZING ACCUMULATORS FOR ISOTHERMAL CONDITIONS

For this condition, Eq. (3) also called the Boyle-Mariotte law can be re-written in terms of V_1 and V_2,

$$V_1 = V_0 (P_0 / P_1) \text{ and } V_2 = V_0 (P_0 / P_2) \quad (5)$$

The difference between V_1 at minimum operating pressure P_1 and V_2 at maximum operating pressure P_2 gives the amount of stored fluid ΔV.

Thus,
$$\Delta V = V_1 - V_2 = V_0 (P_0 / P_1) - V_0 (P_0 / P_2) \quad (6)$$

or
$$\Delta V = V_0 [P_0 / P_1 - P_0 / P_2] \quad (7)$$

or
$$V_0 = \Delta V / [P_0 / P_1 - P_0 / P_2] \quad (8)$$

In the above equations, V_1 and V_2 are nitrogen volumes at pressures P_1 and P_2, V_0 is the nitrogen pre-charge volume at pressure P_0 in liters. It is the maximum volume of gas that can be stored in the accumulator. The size of the accumulator, while conforming to the standard available sizes, should be at least 5–10 percent more than this volume V_0. Any increase in the value of ΔV results in a corresponding increase in the size of the accumulator. Likewise any decrease in the value of P_0 or in the value of $(P_1 - P_2)$ requires an accumulator of higher volume.

3.5 SIZING ACCUMULATORS FOR ADIABATIC CONDITIONS

Starting from the basic formula $P_0 V_0^n = P_1 V_1^n = P_2 V_2^n$ it can be shown that for adiabatic conditions the values of maximum nitrogen volume V_0 at pre-charge pressure P_0 and the stored volume of oil ΔV are given by the following equations.

$$\Delta V = V_0[(P_0/P_1)^{0.7143} - (P_0/P_2)^{0.7143}] \quad (9)$$

$$V_0 = \Delta V/[(P_0/P_1)^{0.7143} - (P_0/P_2)^{0.7143}] \quad (10)$$

Here, the value of the polytropic exponent n is taken equal to 1.4 and therefore $1/n$ becomes = 0.7143. Here again intermediate values can be used for more accurate results.

3.6 SIZING ACCUMULATORS FOR EMERGENCY RESERVE

This is a typical application where both isothermal and adiabatic conditions prevail due to slow storage and quick discharge. For this condition accumulator volume is given by,

$$V_0 = \Delta V (P_2/P_0)/[(P_2/P_1)^{0.7143} - 1] \quad (11)$$

$$\Delta V = V_0 P_0 [(P_2/P_1)^{0.7143} - 1] / P_2 \quad (12)$$

3.7 SIZING ACCUMULATORS FOR PULSATION DAMPING

Pulsation damping is typically an adiabatic condition because both storage and discharge have to be accomplished in a very short time. Because pressure pulsation is a phenomenon associated with piston pumps the stored volume ΔV is a product of the pump displacement q in liters and a constant k which depends on whether the pump is single acting or double acting, and the number of pistons involved. Pressure pulsation is highest in a single acting single piston pump delivering large flows at high pressures. Here the k factor is about 0.69. A single piston double acting pump has the same k factor as that of a double piston single acting pump whose k factor is equal to 0.29. A 3 or 4 piston single acting pump has a k factor of 0.12. For any other pump configuration an average k factor = 0.05 can be taken with reasonable accuracy.

The stored volume $\Delta V = kq$

where q = pump flow rate in (LPM/RPM × number of piston)

$$V_0 = \Delta V / [P_0/P_1]^{0.7143} - [P_0/P_2]^{0.7143} \quad (13)$$

where $P_1 = (p - x)$ and $P_2 = (p + x)$. Here p is the average working pressure in kgf/sq-cm, and x is given by (a × P)/100 kgf/sq-cm.

3.8 SIZING ACCUMULATORS FOR HYDRAULIC LINE SHOCK DAMPING

A suitable accumulator can neutralize water hammering in pipes due to shock waves caused by sudden closure of valves. Typical applications can be found in water, fuel and oil distribution circuits. The volume of the accumulator required to absorb the shock waves is given by,

$$V_0 = 4\ Q.\ P_2.\ (0.0164\ .\ L\ 2t\ )/1000\ .\ (P_2\ 2P_1) \qquad (14)$$

where Q is the flow rate in LPM, P_2 is the maximum pressure in kgf/sq-cm. L is the length of the pipe in meters, t is the acceleration, deceleration or the valve shut off time in seconds and P_1 is the operating pressure with free flow in kgf/sq-cm. In some applications such as hydraulic lift trucks they may be used to absorb hydraulic shocks when the valve shifts or to absorb load induced pressure surges when the truck runs over uneven ground.

3.8.1 Influence of Variations in Temperature on Accumulator Volume

The nitrogen pre-charge pressure in an accumulator is based on the expected maximum rise in the circulating hydraulic oil temperature. This temperature can drop due to changes in environmental factors resulting in a comparable drop in the pre-charge pressure. According to Gay lussac's law this variation in pressure will affect the volume resulting in lower accumulator capacity. It will therefore be necessary to have an accumulator of higher volume so that the useful volume ΔV remains un affected. This correction can be effected by the equation, $V_{oT} = V_0\ (T_2\ /\ T_1)$ where T_2 is the maximum working temperature and T_1 is minimum working temperature in ºK, and V_0 is the volume obtained without accounting for thermal variation and V_{oT} is the new volume capacity of the accumulator in liters.

3.9 INFLUENCE OF PRESSURE ON ACCUMULATOR VOLUME

That, the value of adiabatic index, n lies between 1 and 1.4 is true for perfect gases. But, nitrogen used in accumulators does not behave like a perfect gas when pressure increases and this affects the value of n and consequently the accumulator volume V_0. So, for pressures between 200 kgf/sq-cm to 350 kgf/sq-cm, the value of adiabatic index n may be assumed to lie in the range 1.5 to 1.6.

3.10 SIZING OF ADDITIONAL GAS BOTTLES

Sometimes the size V_0 as determined by the above equations may be more than the available accumulator sizes or the size so determined is too big for accommodation within the frame of the machine or sometimes it may become necessary to maintain small difference between P_1 and $P_{2,}$ which results in a higher stored volume ΔV and a much larger accumulator volume $V_{0.}$ In such cases it is convenient to get required volume by additional bottles.

For example if an application requires 43 liters of fluid to be discharged adiabatically between 70–55 bar, the total volume required would be close to 434 litres. The volume requirement is high because the pressure difference is small. If a 22.5 cm bore piston accumulator can hold 125 liters, auxiliary gas bottles can be installed some distance away to meet the balance of volume requirement.

In applications such as energy reserve, volume compensation and hydraulic line shock damping it is recommended that a higher pre-charge pressure $P_0 = 0.97\ P_1$ be maintained. After obtaining the value of V_0 in the normal way, it must be split in to two portions, one, the minimum indispensable portion which will be contained in the accumulator and the portion which will be contained in additional gas bottles.

WORKED EXAMPLES

Example 1. *A hydraulic cylinder has to move a certain load through a certain distance in one second at a pressure of 140 kgf/sq-cm. An accumulator is integrated into the circuit to provide peak-power. The accumulator is charged for the first 20 seconds and discharges in 2 seconds. The delivery expected from the accumulator is 0.6 liters in 2 seconds as the pressure falls from 250 to 140 kgf/sq-cm. Calculate the accumulator volume. Operating temperature is + 25 C to + 70 C. Also calculate the reduction in input HP due to the accumulator.*

Solution. This is a case of isothermal compression and adiabatic expansion. Equation (11) therefore, is considered here.

$$V_0 = \Delta V\ (P_2/\ P_0)/[\ (P_2/\ P_1)^{0.7143} - 1]$$

Since the maximum system pressure is above 200 kgf/sq-cm, we will have to consider a higher value of 1.6 for adiabatic index *n*. Therefore 1/1.6 = 0.625. Inserting this value in the equation above we have,

$$V_0 = \Delta V\ (P_2/\ P_0)/[\ (P_2/\ P_1)^{0..625} - 1]$$

$$V_0 = 0.6\ (251/127)/\ [(251/141)^{0.625} - 1]$$

$$V_0 = 2.74 \text{ litres.}$$

If we apply the correction for temperature change because the pre-charge pressure P_0 was based on the maximum temperature indicated, we will have the new corrected volume V_{0T} = 2.74 (343/298) = 3.15 litres. *A 3, 5 litre accumulator would effectively serve the purpose.* The delivery from the accumulator is 0.6 liters in 2 seconds or 0.3 liters/sec, or 18 litres per minute.

In the absence of the accumulator the pump has to supply all of this delivery at a pressure of 140 kgf/sq-cm. The HP required would be (18 × 140)/600 = 4.2 kW. Here the efficiency of the pump is not considered. If the accumulator is included in the circuit the pump has to deliver 0.6/20 = 0.03 litres per second or 1.8 liters per minute sufficient to charge the accumulator to a pressure of 250 kgf/sq-cm within the time interval of 20 seconds. Here again the flow required to retract the cylinder is not considered.

The HP requirement in this case would be (1.8 × 250)/600 = 0.75 kW. **The HP saved is (4.2 – 0.75) = 3.45 kW.**

Example 2. *A hydraulic molding press is kept closed at a maximum system pressure of 200 kgf/sq-cm, for a duration of 60 minutes during the curing period. The maximum leakage permitted during this period is 2 cm³/minute and minimum fall in pressure permitted is 198 kgf/sq-cm. Calculate the accumulator volume.*

Solution. This is an application where the accumulator is used as a leakage compensator under isothermal conditions. Therefore Eq. (8) can be used for this application.

$$V_0 = \Delta V / [P_0 / P_1 - P_0 / P_2]$$

Here ΔV = (2 × 60)/1000 = 0.12 litres. P_0 = (0.9 × 198) = 178 kgf/sq-cm. Inserting these values in the above equation, we have,

$$V_0 = 0.12/[179/199 - 179/201]$$

$$V_0 = 13.3 \text{ litres.}$$

A standard 15-litre accumulator would meet the requirement.

Example 3. *A 3 piston single acting pump of flow rate Q = 133 LPM is operating at 20 kgf/sq-cm and at 148 RPM. Working temperature is 40°C. Calculate the accumulator volume needed to limit the remaining pulsation to ± 2.5%.*

Solution. This is a typical condition of pulsation damping in adiabatic phase due to high-speed compression and expansion. For this condition Eq. (13) becomes valid.

$$V_0 = \Delta V /[P_0/ P_1]^{0.7143} - [P_0/ P_2]^{0.7143}$$

Here $\Delta V = k.q$ where k for a three piston single acting arrangement can taken = 0.12. The pump displacement q in liters is given by, q = 133/3 × 148 = 0.3 litres. P_1 = (20 – 0.5) = 19.5 kgf/sq-cm. P_2 = (20 + 0.5) = 20.5

kgf/sq-cm. P_0 = (0.7 × 20) = 14 kgf/sq-cm. Substituting these values in the above equation, we have,

$$V_0 = (0.12 \times 0.3)/[\,(15/19.5)^{0.7143} - (15/20.5)^{0.7143}]$$

$$V_0 = 1.5 \text{ litres.}$$

A 2-litre accumulator would be adequate.

Since all calculations are based on absolute temperature and pressure the temperatures must be expressed in degree Kelvin (º K) that is obtained by adding 273 to the operating temperatures recorded in ºC. Similarly to obtain absolute pressures, 1 kgf/sq-cm is added to the values of P given.

3.11 INSTALLATION AND SERVICING OF ACCUMULATORS

The accumulator must be installed as close as possible to unit it is supposed to serve. There should be sufficient space around both the gas valve and oil entry port. Should be mounted as far as possible vertically with the gas valve on top and with the nameplate details clearly visible.

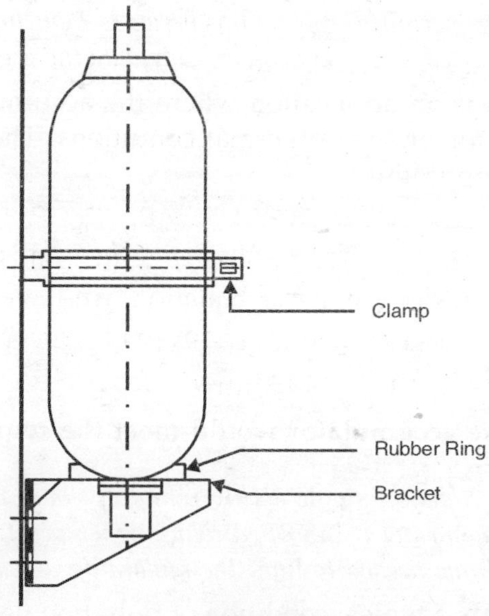

Fig. 3.6

The name plate contains, details such as accumulator capacity, pre-charge pressure and maximum operating pressure. These are vital details for servicing and re-charging purposes and must be protected. An accumulator set consists of a hydro-pneumatic accumulator, a safety and shut-off block, Rear wall bracket, rubber supporting ring, and an accumulator holding clamp.

Optional equipments include a charging unit which will enable nitrogen charging, re-charging and gas pressure alterations. Accumulators unless otherwise instructed by the user are supplied without nitrogen pre-charge.

3.12 INTENSIFIERS

While the accumulator is a power storage device intensifier is a power converter. It uses a large quantity of low-pressure fluid to produce a small quantity of high-pressure fluid. This is possible because of the relationship force = (Pressure × area). The theory involved here is very simple. A pressure of 6 kgf/sq-cm applied over a piston area of 100 sq-cm, creates a force equal to 600 kg. This force when in turn applied over an area 10 sq-cm of the plunger containing a confined fluid will generate a pressure equal to 600/10 = 60 kgf/sq-cm. This high-pressure fluid when admitted in to the chambers underneath a piston of area 100 sq-cm, can lift a load (60 × 100) = 6000 kgf or 6 tonnes. The advantage is obvious.

By merely applying a pressure of 6 kgf/sq-cm, which even a small air compressor commonly found in any workshop can develop we are obtaining force of 6 tonnes without the need for a hydraulic power pack. Pressure increase is in inverse proportion to the area ratios.

The volume of high-pressure fluid that is discharged is proportionately less compared to the input volume. This is the basic disadvantage of the intensifiers. They are therefore only used in low volume, high-pressure 'one shot' applications such as job holding, cutting, crimping, or in hydraulic power tools where the piston stroke is very small.

In applications where the piston stroke is long, compressed air is used to extend the cylinder through most part of its stroke with intensifier stepping in to supply high-pressure fluid only after work resistance is encountered. A small compact intensifier can be built to develop pressures as high as 700 kgf/sq-cm, using a small low-pressure pump or tapping a line from the existing circuit. The ratio of intensification is the same as the ratio of two piston areas. The intensification can be as high as 10 : 1. This means a 6 bar air pressure can be converted in to a 60 bar oil pressure, in an air-to-oil intensifier. Air-to-oil, oil-to-oil, and air-to-air are three kinds of intensifiers in use. Because air is compressible air-to-air intensifiers are less predictable than other types. If shop air is readily available, air to oil intensifiers offer an economical choice. Where very high pressures are required and where low-pressure hydraulic supply is available then oil-oil type is the best choice.

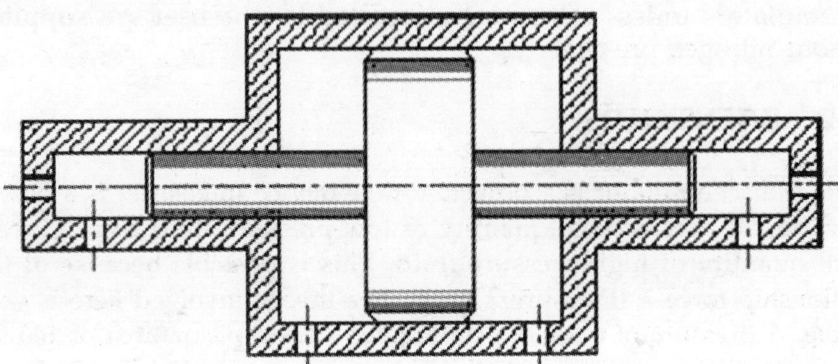

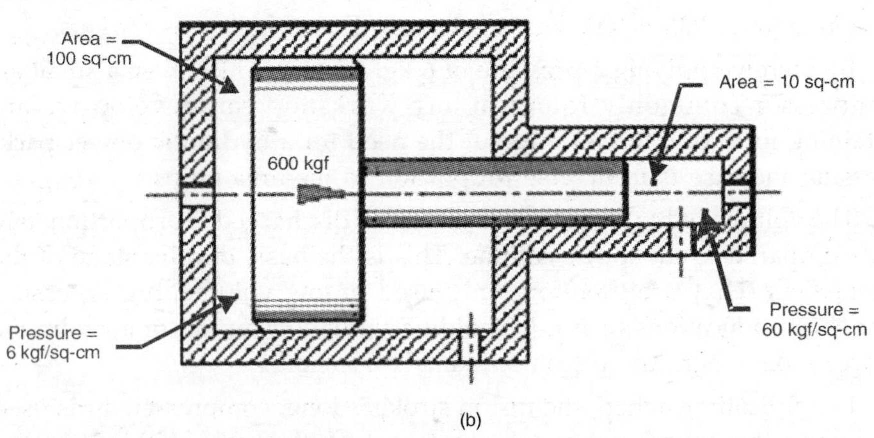

Fig. 3.7. *(a) Double acting intensifier; (b) Single acting intensifier.*

Intensifiers are also called 'pressure boosters' and are commonly used to deliver high-pressure fluid only during the last portion of the cylinder stroke. The normal circuit extends and retracts the actuator. Intensifier stroke L required for such an application is given by $L = [(V_P + V_0)/A_R] + S$, where V_p is the volume required to move the cylinder through the high pressure portion of the stroke, V_0 is volume loss due to compressibility of the hydraulic oil, A_R is the intensifier ram area and S is intensifier ram travel. One percent oil volume loss due to oil compressibility may be assumed for every 70 kgf/sq-cm.

Intensifiers have a large and a small piston mounted on a common piston rod and housed in bores of appropriate diameters. They can be single acting or double acting. Single acting intensifiers need some mechanical means to return the piston to its original position after each operation. Double acting intensifiers require valves to alternately change the direction of oil flow.

Since the valves need to be placed only in the low-pressure side they do not present a problem either in terms of cost or availability. Double rod end intensifiers supply intensified fluid during both the forward and return motion of the piston to increase fluid output.

Intensifiers deliver high-pressure fluid at the point application and to the extent required. This eliminates the need for a pressure relief valve in turn eliminating heating up of oil. They offer a simple and compact alternative solution to the need for high-pressure pumps.

4

POWER TRASMISSION: HYDRAULIC FLUID

Oil in hydraulic system performs the dual function of transmission of power and lubrication of rotating or sliding parts. It constitutes a vital factor in the hydraulic systems and therefore proper selection of oil is required for satisfactory life and operation of the system components, especially the hydraulic pumps and motors. Any oil selected for use with pumps or motors is acceptable for use with valves. Mineral or petroleum oil of various grades, are the most common power transmission members in hydraulic machines and machine tools.

Some of the factors, which are of vital importance in the selection of oil for use as power transmission members are discussed as below.

1. Viscosity. In simple terms viscosity is the measure of the resistance to flow of any fluid. If a fluid flows easily it is less viscous and vice versa. As compared to hydraulic oil, water is less viscous since it offers least resistance to flow. This is one of the reasons why water as such is not acceptable as hydraulic fluid medium since its low viscosity presents sealing problems even at moderate pressures. Highly viscous fluids on the other hand offer very high resistance to flow, which in turn will result in wasted horsepower. The viscosity of the fluid selected ultimately, for a certain application is usually a compromise between these two extreme conditions.

Values of viscosity of hydraulic oil recommended for most applications generally falls in the range of 32 to 46 centistokes (cSt) at 38° C. Also less viscous oil (16–32 cSt) or more viscous oil (46–72 cSt) can be used, keeping in mind that less viscous oils have to be used in low pressure systems to limit the pressure drop and more viscous ones in high pressure systems to limit the leakage losses.

Apart from centistokes other units of measurement used are, say bolt universal seconds (SUS) and degrees angler. Viscometer is the instrument that is used to measure the viscosity of the fluid.

2. Viscosity Index. Viscosity index on the other hand, indicates how the fluid viscosity changes with increase in temperature. Ideally, the fluid should have the same viscosity at very low temperatures as at very high temperatures. But in reality this is not so. As the oil temperature increases the viscosity of most hydrocarbon fluids decrease. This variation can affect work performance especially in grinding machines or in tracer controlled system. Modern hydraulic oils, which have a paraffin base, have a viscosity index of 90–105 as compared to oils with naphthenic base, which have an index of only 35.

Higher the viscosity index, higher will be the temperature required to bring about changes in the viscosity of the fluid. To obtain optimum service life from the oil the normally recommended operating temperature to be maintained is between 40° C to 65° C. If the oil temperature is maintained beyond this limit consistently it will cause rapid seal wear out, increased leakage losses due to thinning of oil and malfunction of hydraulic solenoid valve. A high viscosity index becomes important in applications subjected to wide temperature range, mobile hydraulic systems used out doors, hydraulic systems in use during winter in places where the temperatures can go below zero and the like. Special additives are sometime used to improve the viscosity index of most fluids.

3. Pour Point. This is the lowest temperature at which the oil flows, when chilled under specified test conditions. This parameter becomes important, only if the system is regularly exposed to low ambient temperatures.

4. Compressibility. Hydraulic oil is practically incompressible at low pressures but may become somewhat compressible at higher pressures. This factor has marked effect on the dynamic rigidity of the system. The compressibility of the fluid will depend on its bulk modulus. Hydraulic oil will compress at about 70 kgf/cm^2 by about one half of 1% of its volume, which is about 0.5%of its original volume. At 700 kgf/cm^2 it will compress by 4% of its original volume. Compressibility or bulk modulus has its greatest effect on performance in servo applications. Because compressibility increases with pressure and temperature, it becomes an important factor in high-pressure applications. In positive displacement pumps, the effect of bulk modulus shows up as a loss in volume because few actuators recover the compressive energy in the fluid.

5. Stability. Fluids used in hydraulic systems as power transmission and lubricating medium must retain their properties for a long time. But mechanical stresses due to flow and cavitation can shear polymer chains and thus reduce viscosity.

Oxidation and hydrolysis can cause chemical changes resulting in formation of volatile components, insoluble materials and corrosive products. Heat can also destroy a fluid. Higher the operating temperature, shorter will be the life of the fluid. An old rule of thumb states that for every 10º C rise, the rate of oxidation doubles. Thus a fluid used at 110º C has about half the life as at 100º C. Corrosive agents usually form due to thermal decomposition or during hydrolysis. These corrosives are usually acidic but not all acidic materials are corrosive. Corrosion increases leakage by increasing the clearances between closely fitted parts. If pitting corrosion occurs there can be substantial localized loss of strength.

6. Lubricating Ability. This is an important quality in hydraulic fluids since the fluid must lubricate moving parts of the system to minimize wear. Most hydrocarbon fluids have good lubrication properties. But in cases where pumps and motors impose severe load carrying requirements on the oil then the fluid should be fortified by anti wear additives. Where boundary layer lubrication conditions prevail, glycol based fluids are satisfactory. They are stable in hydraulic service and are unaffected by high rates of shear.

7. Resistance to Foaming. Fluid in a hydraulic system always contains dissolved air. This air tends to increase the compressibility of the fluid, making the system elastic, noisy and erratic. Frequent *compression of this air generates heat and can increase oxidation. Volumetric* efficiency of the pump is reduced because air bubbles in the oil on the inlet side expand as the oil enters the pump. When the bubbles subsequently collapse on the discharge side, the damage caused is similar to cavitation corrosion. Many high quality fluids contain anti-foaming additives that readily releases air.

8. Materials Compatibility. Natural rubber is not oil resistant and should not be used in hydraulic systems using petroleum oils. Synthetic rubbers vary widely in their behaviour when exposed to different fluids. In contact with a certain fluid, some are unaffected but others swell, shrink or otherwise deteriorate. Nitrile rubber is the most common material used in hydraulic systems. For special applications, the choices are, neoprene, silicones and fluorocarbons.

9. Sound Level. Very high viscosities at start up temperatures induce pump noise due to cavitation. Even moderately viscous fluids impede the release of entrained air, thus increasing the sound levels. Contaminated fluid can cause rapid wear out of pump parts, which may result in increased sound levels.

The oil change must be made as a rule every 2000 hours. The lasting time indicated is approximate and must be evaluated based on the physical and chemical properties of the oil. A number of techniques have been developed to evaluate the condition of the oil, circulating in the system before a change is effected.

Patch testing is the basic field test. In this test the fluid sample is drawn through a 0.8-micrometer membrane patch, which leaves the contamination on the surface. After the patch is rinsed with solvent to remove excess fluid, the membrane is viewed under 100-x magnification. Patch test provides a good qualitative impression of the contamination level of the fluid, because it provides a rough estimate of the percent of metal particles in the fluid. This helps to determine whether the dirt source is wear or ingression. Patch test does not quantify dirt and is subject to error.

Gravimetric analysis is a measure of the weight of solid material present in a specific volume of fluid. A known volume of fluid is passed through a 0.8-micrometer analysis membrane. The weight of solid contaminant is reported as milligrams of contaminant per liter of fluid. This method is easy to perform and is reproducible.

Table. 4.1. *Equivalent Chart for Hydraulic Oils*

Class	*Description*	*ISO Grade VG ***	*B.P*	*H.P*	*IOC*	*ESSO*	*Shell*	*Mobil*
HA	Highly refined mineral oil with anti-rust, anti- foam, anti-oxidation and anti-wear properties. Has high viscosity index	32 46*** 68	Hydrol	Enklov	Servo	Nuto	Tellus	DTE 24 DTE 25 DTE 26

** Corresponds to Mid Point Value of Kinematic Viscosity in CST at 40º C with a Tolerance ± 10%

*** Preferred Oil For High Pressure Hydraulic Applications

BP–Bharath Petroleum, HP–Hindustan Petroleum, IOC–Indian Oil Company

4.1 RE-REFINED HYDRAULIC OIL

A barrel of virgin hydraulic oil in India costs around Rs.10,000 that is, approximately Rs. 50 a liter or nearly a dollar a liter at the exchange rate prevailing today. It costs more than a litre of gasoline and so the cost can be considered exorbitant. Reclamation of used hydraulic oil therefore becomes sound economics for the nation, for the business or the concerned individual.

Because oil never wears out but only gets contaminated, re-cycling of used hydraulic oil is both feasible and profitable.

Visual inspection of hydraulic oil should be sufficient to determine the quality of the oil. Darkened appearance, foaming, water content, viscosity changes, existence of solid contaminants, are the characteristics that helps conclude the extent contamination of the oil. Re-refining of hydraulic oil should be considered for degraded oils that cannot be improved by settling and filtration. Re-refining involves dehydration of the contaminated oil, acid treatment to remove oxidation products, neutralisation to restore colour followed by filtration and blending with additives to obtain specific properties. If strict quality standards are maintained at the re-refining plant to ensure a final product that is nearly as good as the original oil, use of re-refined oil should be encouraged. It would be ideal if consumers by an arrangement with the supplier or dealer can exchange their contaminated hydraulic oil with re-refined ones of the same grade for a reasonable difference in prices.

4.2 NON- HYDROCARBON FLUIDS

The mineral oils have quite a low ignition and combustion temperatures. Therefore, in cases where either because of higher operating temperatures or due to risk involved when the fluid leaks to surroundings, which can cause fire hazard, non-petroleum fluids are used. They are invariable chosen when work environment demands a fluid that does not spoil food related products or pollute a river. There are two types of non-hydrocarbon fluids.

Phosphate Ester Fluids. The phosphate ester fluids have high resistance to combustion and lower tendency for flame propagation. They have good lubricating and rust inhibiting properties. They have lower viscosity index, high specific weight. But they are non-compatible with rubber and are also toxic. Therefore it becomes important when using phosphate ester fluids, to use seals made of elastomers such as viton or PTFE. The use of rubber hoses is not recommended. The phosphate ester fluids have wide operating temperature ranges (0–80ºC) and an optimum resistance to aging. They do not require special maintenance. They can be used with piston/vane pumps or motors at full catalogue ratings. The minimum in let pressure requirement is about 35% higher than petroleum based fluids.

Water-Glycol Fluids. These are compounds of water (40–50 %) the rest being ethylene or propylene glycol. Besides they have additives to improve viscosity, prevent rust foaming, corrosion and improve lubrication. The combustion resistance is because of its water content. These fluids have high viscosity index (160–200) , fairly good lubricating and rust inhibiting property. They are not compatible with most of the seal material and paints.

The recommended seal material is Buna-N or viton-A. The maximum operating pressures are in the range of 100–120 kgf/cm^2. Gear pumps cannot be used with water glycol solution. Piston and vane pumps cannot be run in excess of 1000–1200 rpm due to poor lubricity of these fluids. The minimum inlet pressure should be around 13 psia. ISO 7745 standards give different grades of fire resistant fluids available. Table 4.1 depicts the commercial grades of hydraulic oil normally used in oil hydraulic systems.

4.3 CONTAMINATION CONTROL

Contamination control as applied to oil hydraulic system means maintaining the purity and virginity of the power transmission medium, usually the hydraulic oil. The benefits are two fold. (1) Premature discard of the oil is prevented. (2) Frequent failure and troubleshooting resulting in machine down time or one time catastrophic failure of the machine resulting in machine discard is prevented. The cost of poor contamination control should therefore be obvious.

Despite the threat it poses only a few system designers and end users do appreciate the importance of contamination control in hydraulic systems. Lack of appreciation of this vital fact both by the system designer and the end user has been the single most important reason, for frequent failure of the otherwise well conceived and executed hydraulic systems.

Four different sources of contamination have been identified as follows.

1. Those generated within the system. These are ferrous fines due to abrasive wear caused by silt getting lodged in the working clearances of components. Ferrous fines may also originate due to erosion of the metal caused by high velocity metal particles. Chemical reactions due to the presence of water in the oil, or pitting due to cavitations problem in the pump that initiate corrosive wear of metal surfaces.

2. Built-in contaminants. These are contaminants lodged in the crevices of hydraulic components such as actuators, valves, reservoirs, pipe lines, pumps etc. which gets dislodged on application of high pressure.

3. Normally ingressed contaminants. These are dust particles gaining entry into hydraulic system through the breathers, tank openings, piston rods etc.

4. Contaminants that gain entry during maintenance. This can occur either during topping up or during assembly and disassembly.

Hydraulic oil may be considered as contaminated, if it contains particulate matter foreign to its composition such as, ferrous chips and silt.

Free or dissolved water and free air are also considered as contaminants. In average hydraulic systems operating at pressures between 140–280 kgf/sq-cm, where 'chip' control will assure satisfactory performance, the expected level of cleanliness is 17/14 conforming to ISO 4406. In more critical applications operating at higher pressures, using servo valves or proportional valves, where 'silt' control will be required, the expected level is 14/11. Contamination level is given in Table 4.2.

Table. 4.2. *Contamination Chart.*

Standard		*Number of Particles per 100 ML.*				
SAE 749D	*ISO 4406*	*Micron Range mm*				
Contamination Level		*5 to 15*	*15 to 25*	*25 to 50*	*50 to 100*	*> 100*
1	13/10	4000	712	126	22	4
2	14/11	8000	1425	253	45	8
3	15/12	16000	2800	506	90	16
4	16/13	32000	5700	1012	180	32
5	17/14	64000	11400	2000	360	64
6	18/15	128000	22800	4100	720	128
7	19/16	256000	45600	8100	1440	256
8	20/17	512000	91200	16200	2800	512
9	21/18	1000000	182000	32400	5800	1024

Poorly controlled contamination can cause frequent system failures due to jamming of relative surfaces which work with critical clearances leading to malfunction which can be rectified or slow deterioration of the system resulting in premature failure of a component or sudden catastrophic failure. In terms of loss of productivity and costs of rectification/replacement of parts it is a high price to pay. This cost will surely justify the cost of effective contamination control.

Suction strainers / inlet filters, pressure line filters and return line filters and by- pass filters are the various components usually installed in a hydraulic system to trap contaminants of the type narrated above. Whether a suction strainer, a return line filter or a pressure line filter, they all perform the same task that of trapping the contaminants up stream so that the components down stream are protected. The only difference being in the size and number of contaminants that can get past them.

This is indicated by their β ratio, which is the ratio of the number of particles in a certain size range upstream of the filter to the number of particles in that size range down stream.

For instance β_{10} = 2 means that in the size range of particles larger than 10 mm, there are twice the number of particles, upstream of the filter as down stream. This in effect means a filter rated β_{10} =2 retains half the contaminant particles larger than 10 μm.

A typical filter unit consists of a body with integral bypass valve placed in between the inlet and outlet ports. Bypass valves are designed to open and allow free passage of the fluid if the filter element gets clogged due to poor maintenance. The body and the housing for the filter element are held together either by the threads formed on the body and the housing or held firmly by a central fixing screw. With the help of O-rings at appropriate places a leak proof enclosure is formed. Filter units must be mounted vertically with the filter housing facing down to facilitate easy removal of the element for cleaning. Filter elements made of stainless steel straight wire mesh have coarse holes. A 100-mesh screen provides a nominal filtration of 150 μm. A 200-mesh screen will provide a nominal filtration of 75 μm. A sintered woven wire mesh can provide upto 10 μm. These are basically metal chip control filters.

These filters can be reused any number of times by cleaning them with a solvent and by blowing air from inside to out side. Silt control filter elements made of fiber glass or cotton wound can provide filtration from 1—5 μm. The elements cannot be reused but have to be replaced. Suction strainer is so called because it is a coarse filter and is used in the pump suction line. It is usually without its housing and is always fully immersed in oil. Its micronic rating is usually 150 μm or 100 meshes. This means, it can trap particles of size more than 0.150 μm. In the suction line micronic rating less than this is not recommended as it might starve the pump of hydraulic oil, resulting in increased pump noise and consequent cavitations damage to the pump. Suction strainers are specified on the basis of their micronic rating and fluid handling capabilities, which depends on the mesh area. Strainers capable of handling flows about three to five times the pump discharge are selected.

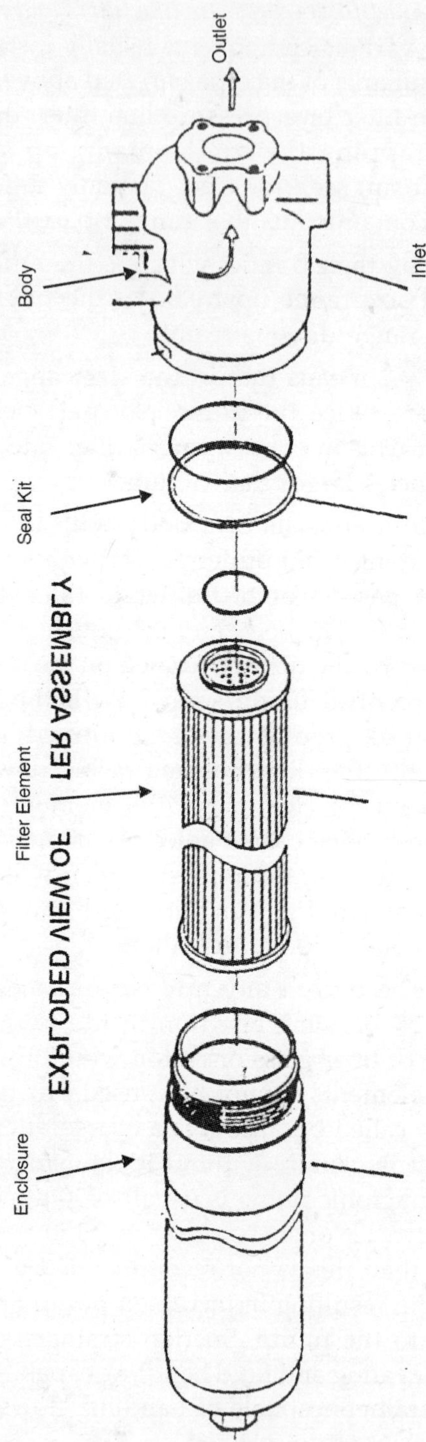

Fig. 4.1. *Exploded view of filter assembly.*

Sometimes, two strainers are used in parallel, if one of sufficient capacity is not available. Periodic removal and cleaning of the strainer is a task that needs to be performed without neglect to protect the pump from premature and permanent damage. This is because, unlike filters,

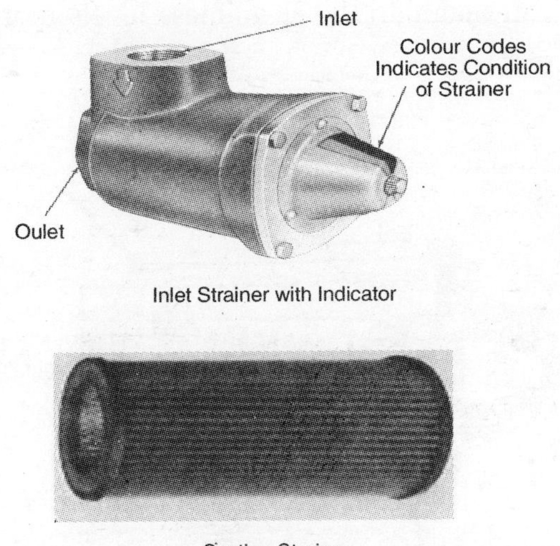

Inlet Strainer with Indicator

Suction Strainer

Fig. 4.2

strainers do not have by pass facility when fully clogged. To overcome this problem, indicator filters, which signal the operator when cleaning is required, have been developed for the inlet lines. These, also incorporate by pass facility. But their initial and installation costs are higher compared to the simple suction strainer. Return line filters are placed in the line that carries the fluid back to tank. A properly sized return line filter is most popular in industry environment. Their micronic rating is 15 μm. This means the filter would trap contaminants of size more than 0.015 mm. They have a bypass facility built in. Since they are mounted above the tank, it is comparatively easy to remove and clean/ replace the filter element. Return line filters are specified on the basis of their micronic rating, flow handling capabilities and the differential pressure between inlet and outlet ports. They are rated for a maximum pressure of 10 kgf/sq-cm.

Pressure line filters are located between the pump and the control valves. Being in the pressure line, they are subjected to the system working pressure and therefore, the filter housing must be designed to withstand the forces involved. Their micronic rating is between 1–5 μm. They are specified on the basis of their working pressure, micronic rating and flow handling capabilities. Both by pass and clog indicating facility is mandatory.

Filters may also be located in a by pass line in the hydraulic circuit in which case they are called 'bypass filters'. Only a small portion of the total pump flow is allowed to pass through them, the rest of the flow being fed directly to the system components. In this case, a small filter of appropriate mesh can maintain the cleanliness levels more economically and efficiently.

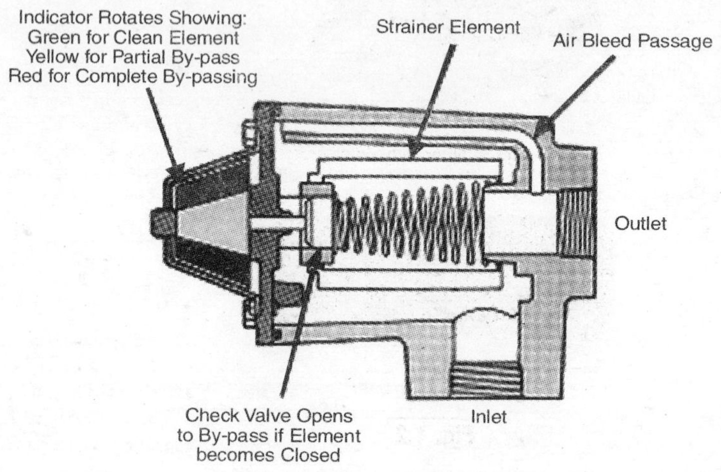

Fig. 4.3. *Pressure line filter.*

While selecting a filter for the hydraulic unit the factors that needs to be considered are, the working pressure, pump discharge, duty cycle, sensitivity of the components used and productivity expected of the machine.

The following procedures are recommended.

1. Suction strainers/Inlet line filters are required to protect the pump from the debris in the reservoir. But they are not enough to effectively control contamination. An additional return line filter of 15 μm is required to ensure a cleanliness level of 17/14.
2. The manufacturer's recommendations regarding the cleanliness levels to be maintained for satisfactory performance of their product must be the criteria for selection and location of filters.
3. Pressure/return line filters must be selected based on their capability to handle the pump discharge at a pressure drop no more than 0.5 kgf/sq-cm.

4.4 FIRST LINE FILTERS

1. Magnetic Tank Cleaners. These are a multi magnet system consisting of three or more round ceramic magnets held together by a central rod, which can be fixed to the tank such that the magnets remain immersed in oil. These magnets attract ferrous fines and paramagnetic particles. They help isolate contaminants to some extent. Flow disturbances or saturation of the magnets with ferrous fines can cause the contaminants to get dislodged.

2. Tank Filler-breather Units. These are screwed on to an opening in the tank top or sides. They serve as oil cap, dirt screen and air filter. They are provided with a 40 mm mesh and have to be selected based on the rate of airflow required.

3. Portable Filter and Transfer Unit. These are trolley mounted units used in-plant or on-site if single-phase power is available. They can filter the existing oil in the tank, or flush it out and transfer fresh oil from the container. They can provide a cleanliness level of 3–10 mm.

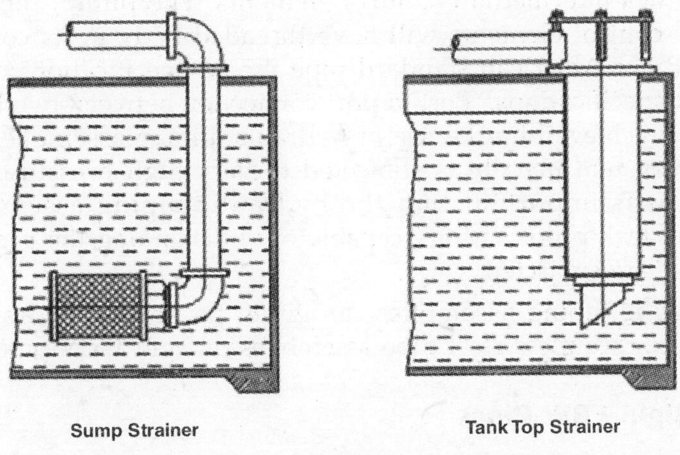

Fig. 4.4

It is generally preferable to use tank top mounted inlet strainers instead of oil immersed sump strainers. They are difficult to service without shutting down the machine for element inspection/change. Tank top strainers permit easy access to filter element from out side. They are available as combined filter and breather units with standard integral by pass set to 1.5 kgf/sq-cm.

Any strategy adapted to control the deleterious effects of contamination should first bring about an awareness of its importance in concerned people, such as the assembly technicians, the maintenance technicians etc. This will make sure, the hydraulic oil, the oil reservoir and other hydraulic components are thoroughly clean in the first place. Using an ultra fine mesh filter in an already contaminated oil, would only compound the problems of contamination since the filter will go into the by pass mode within hours after start up. Also the filter chosen must be easily serviceable. The housing should open quickly with minimum of tools for cleaning/change of element etc.

Inlet line filters must be used in place of submerged inlet strainers, where it is not possible to provide easy access for their removal and cleaning, without draining the tank. Having done this, instructions must be displayed on the machine, in the instruction manual etc., regarding the frequency of servicing required. The above safeguards, if followed meticulously would solve the problems of contamination to a large extent.

4.5 FLUID CONDUCTORS

They conduct hydraulic oil from the pump to the actuator through various other intermediate control elements. The pump, the actuator and other control elements will have threaded body ports conforming to either BS or American standard pipe thread specifications or to SAE 2 or 4 bolt specifications. Port to port connection between two hydraulic elements is achieved by the use of hydraulic piping assemblies. Since a hydraulic system usually is subjected to very high pressures (50–500 kgf/cm^2) it is important that the hydraulic piping assemblies are completely leak proof and are capable of withstanding the pressures to which they are subjected.

Hydraulic piping assemblies can either be flexible hydraulic hose assemblies or seamless steel tube assemblies or rigid ERW pipes.

4.5.1 Rigid ERW Pipes

These are invariably used on the suction side and low-pressure side and where large volumes of fluid must be transferred such as in pre-fill valve systems. Where the pipe diameter is small they are normally connected directly to pipe thread connections without the use of end fittings. Teflon tapes are usually wound on taper threads formed on these pipes before they are tightened to mating parts. When tightened properly they form joints sufficient to prevent entry of air and leakage of oil at low pressures. Where the pipe diameters are large, flanged joints with intermediate gaskets are used to connect these pipes on to their mating parts. They are rigid, in the sense that they cannot be bent and therefore they can only form straight connections.

Pipes are classified based on their nominal size and schedule number. Nominal size indicates, only the thread size for connections. The outside diameter is always bigger than the nominal size. Schedule numbers indicates the wall thickness of the pipe. A pipe may have one nominal size but different schedule numbers. A variety of fittings such as Tee, 90-degree elbows, parallel couplings, reducer couplings etc., are available to complete the pipe connections.

Flexible Hose Assemblies. It consists of a flexible hose, with appropriate end fittings or inserts crimped to the hose on either end. They are widely used in applications where the lines must flex and bend or in applications where the connected part is not fixed rigidly and is free to move or oscillate with the load. A flexible hose can be as simple as a plastic tube with either plastic or metal end fittings. Plastic tubes are never used in hydraulic system except in applications, which calls for mere transfer of fluid at low temperatures, from one location to another. They cost far less than other hose assemblies, reduce noise, and withstand assault by a wide range of corrosive fluids. Because their internal surfaces are smooth they offer low resistance to flow. They are never good for pressures beyond 15 to 20 kgf/square centimeter and temperatures beyond 70–80ºC. PVC, nylon, polyethylene, and PTFE are the most common plastic material used for these hoses.

Hydraulic Hose Assembly. Hydraulic systems operate at pressures beyond 50 kgf/sq-cm. The application therefore calls for high-pressure hydraulic hoses, which are a multi-layer construction with the outer and the inner layer made of synthetic rubber. The intermediate layers are either fabric, wire braided or spiral wound reinforcements. Spiral wire reinforcement provides greater flexibility, can withstand higher pressures but they are more expensive than simple wire braided hoses.

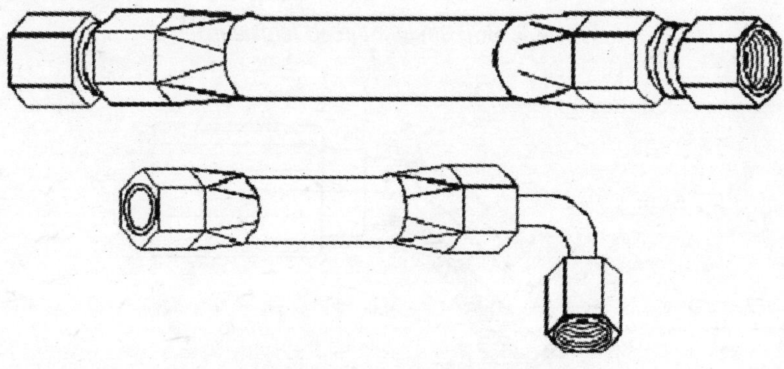

Fig. 4.5. *Hydraulic hose assembly.*

Both wire braided and fabric reinforcements are sometimes used alternately up to six layers in certain applications. Hose construction has been standardized under SAE 100R series ranging from SAE 100R1-SAE 100R14. These hoses can withstand temperatures from –40º C to + 120º C. The working pressure depends on the hose inside diameter. A SAE 100R1 hose which is a single wire braided construction, has a pressure rating of 200 kgf/sq-cm for a 5 mm ID hose where as for a 50 mm ID hose of the same construction, the pressure rating is just about 25 kgf/sq-cm. Working pressure also depends on the bend radius of the hose and service life. Hose manufacturers generally give data pertaining to the minimum bend radii for various hose IDs. Bend radius tighter than recommended values over stress the reinforcing braid and shorten hose life besides constricting the flow passages. Under pressure, a hose assembly may change in length up to 4%. So sufficient slack should be provided to allow for shrinkage or expansion and to accommodate pressure surges.

End fittings for the hoses are usually carbon steel inserts, which are crimped or swaged on to the hose at either end to form the hose assembly. These inserts are available in a wide variety of configurations to suit every application. The most common configurations are, the BSP straight female insert, BSP 45° swept elbow insert, 90° BSP swept elbow inserts. These inserts are also available to JIC and SAE standards. For very high-pressure applications, inserts with O-ring seal are preferred. If the hose and coupling are properly selected and if the crimping of the assembly has been effective, the hose assemblies do last for a long time.

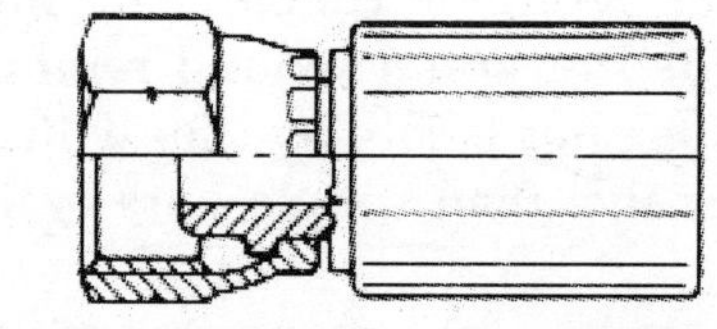

Fig. 4.6. *BSP straight thread female insert.*

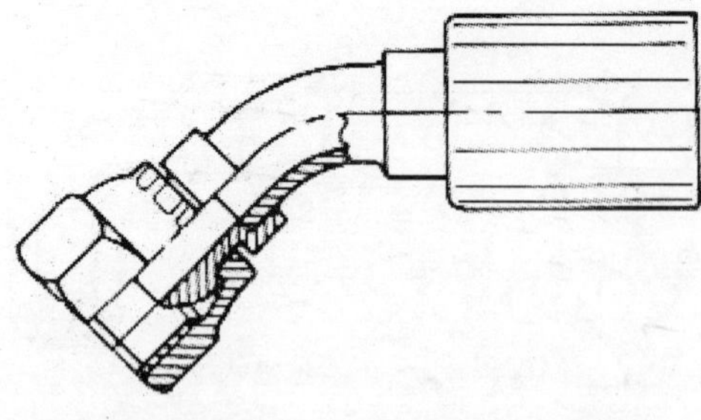

Fig. 4.7. *BSP 45 degree female elbow insert.*

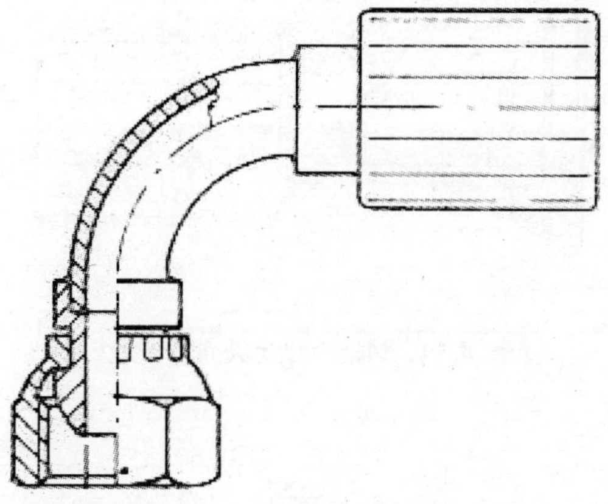

Fig. 4.8. *BSP 90 degree female elbow insert.*

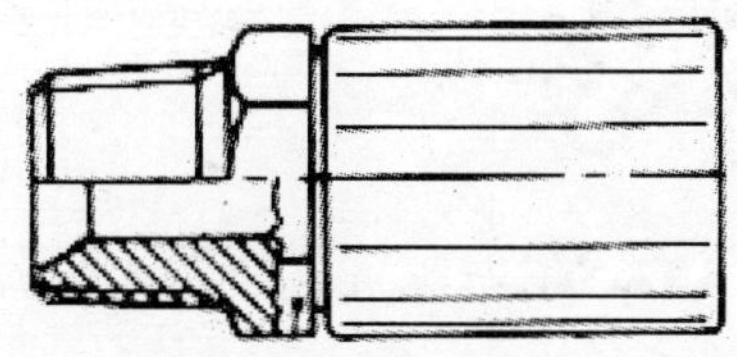

Fig. 4.9. *BSP straight male elbow insert.*

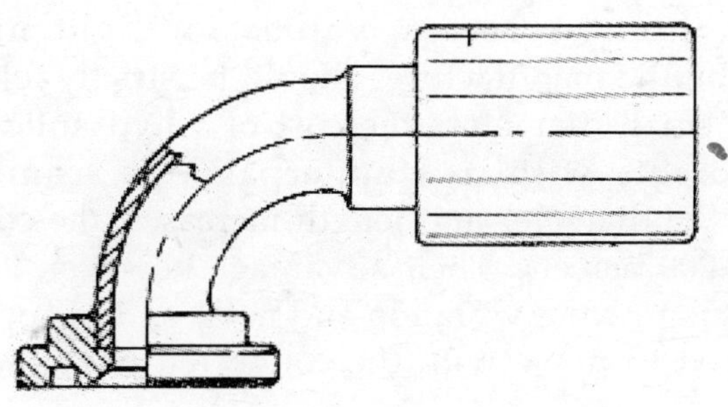

Fig. 4.10. *SAE 90 degree flange elbow insert.*

The hose assembly is designed to bend or flex and not twist. The hose gets twisted usually at the time of tightening the end fittings on to their components. Hose fittings unlike ferrule fittings should not be tightened too hard, as this will tend to twist the hose and damage the crimped end fittings. Extension levers on spanners should not be used to

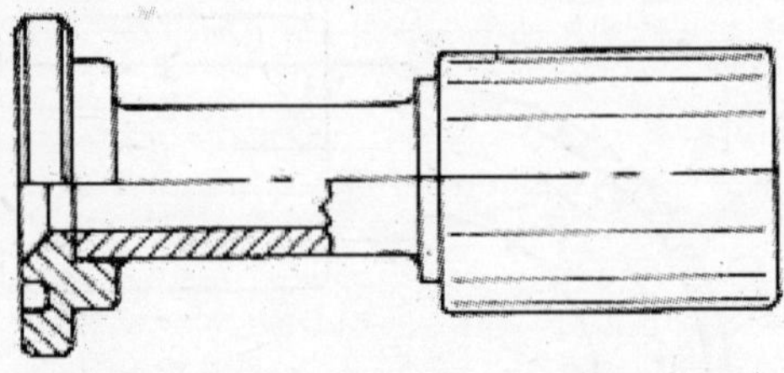

Fig. 4.11. *SAE straight flange insert.*

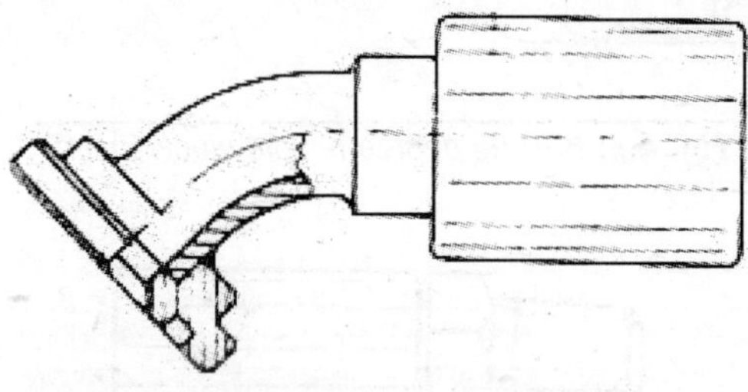

Fig. 4.12. *SAE 45 degrees elbow flnge insert.*

tighten the hose assembly on to their fittings. Clamping of hoses at certain places may be desirable especially if the hoses are long, to avoid chaffing or tangling with moving parts. If at all failure occurs, it will be due to over torqued threads. Torque values for tightening couplings recommended by the manufacturer should be strictly followed. In short lengths and in small diameters the cost of a hydraulic hose assembly compares favorably with its counterpart, the seamless steel tube assemblies. As the diameter and length increases, the cost increase will be between 50–200 percent. Their advantage however, is obvious. They are easy to install, prevent vibration and noise and in applications where they are required to move with the connected part, they are the only choice.

4.6 SEAMLESS STEEL TUBE ASSEMBLIES

Cold drawn seamless steel tubes are available up to 30 mm diameter. Beyond this range, hot drawn steel tubes have to be used. Never use ERW tubes in the hydraulic systems on the pressure side. ERW tubes are acceptable only on the suction side. Seamless carbon steel tubes must be purchased in fully annealed and normalised condition, as otherwise they would develop cracks while bending. Tubes up to 20 mm in diameter are bent using a bending fixture in cold condition without filling them up with sand or other supporting material. Tubes beyond 20 mm are usually bent after filling them up with sand. Very large diameter tubes are bent both in hot and filled up condition. Test pressure for these tubes is usually twice the working pressure. Seamless steel pipes are classified as light and heavy duty based on their wall thickness. The pressure ratings vary accordingly from about 40 to 400 kgf/cm^2.

The wall thickness of the pipe is a function of the system pressure and is determined using the basic equation $T= pdS/2f$ where, T is pipe wall thickness required in centimeters, p is the maximum system working pressure in kgf/sq-cm. d is the inside diameter of the pipe in centimeters, S is the factor of safety and f is the material tensile strength in kgf/sq-cm. The recommended safety factor is between 4 and 8 which depends on the operating pressure, expected mechanical abuse, and shock and vibration levels.

4.7 END FITTINGS FOR STEEL TUBES

37-degree bite-less flared tube fittings and bite-type flare-less tube fittings and seal lock fittings are the types most commonly used in hydraulic systems. In bite less flared tube fittings, the ends of the seamless pipes needs to be flared to 37 degrees to enable the flared tube fittings to form a leak proof joint. Since flaring of thick walled steel tube is difficult operation, the use of bite less flared tube fittings is confined to soft materials like aluminum, copper and thin walled steel tubes and therefore are not preferred on high pressure hydraulic lines. The most commonly used fitting is *the flare less bite type* which, is a unique three piece construction, consisting of a body, a special case hardened carbon steel ferrule and a nut. The ferrule acts like a wedge when tightened over the body and forms a seal between itself and the body, while the cutting edge of the ferrule, bites in to the tube wall to form another seal. Because the ferrule is self-centering, it gives an even bite around the entire tube periphery. The extent of the bite is visible, thus confirming the grip. These fittings provide an excellent seal, and the nut will not loosen even under vibration and strain. The fitting can be dismantled and assembled any number of times, without loss of strength and sealing qualities. The nut is slowly tightened by hand allowing the ferrule to slide freely along the pipe guided by its body taper. The locking section of the ferrule wedges itself, between the body and the pipe surface.

The pipe is now held firmly in position abutting the body, ready for full tightening. Upon further tightening, the ferrule bites perceptibly in to the pipe surface thus providing a leak proof joint.

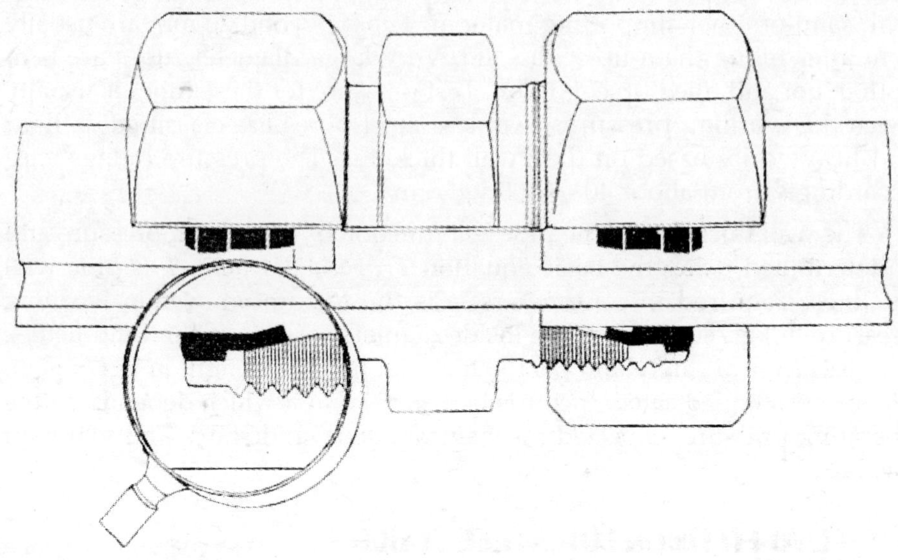

Fig. 4.13. *A flareless bite type ferrule fitting.*

These couplings are available in various configurations to suit every need. A few of them are depicted below.

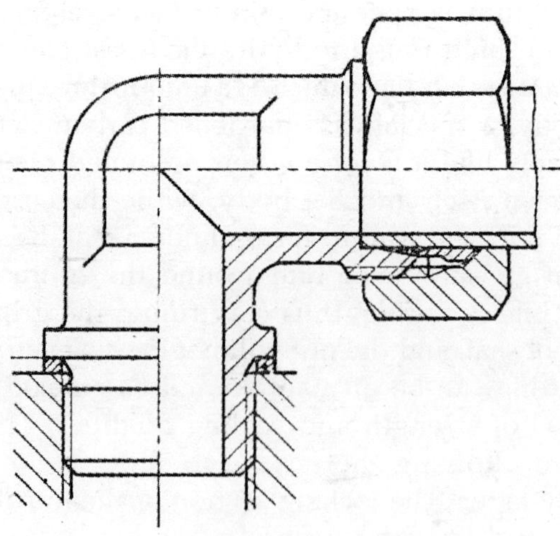

Fig. 4.14. *(a) Parallel male stud elbow.*

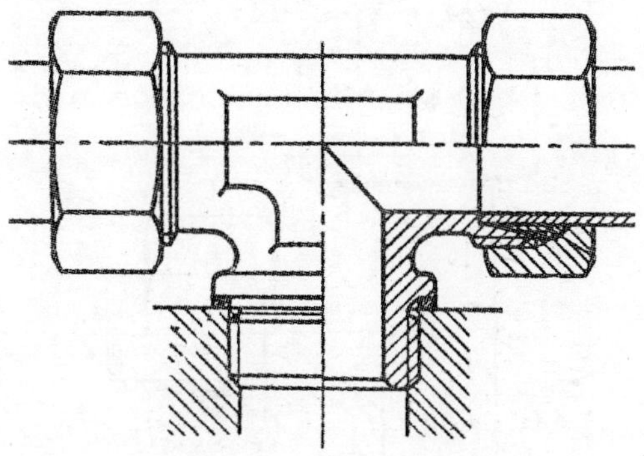

Fig. 4.14. *(b) Parallel male stud 'T'.*

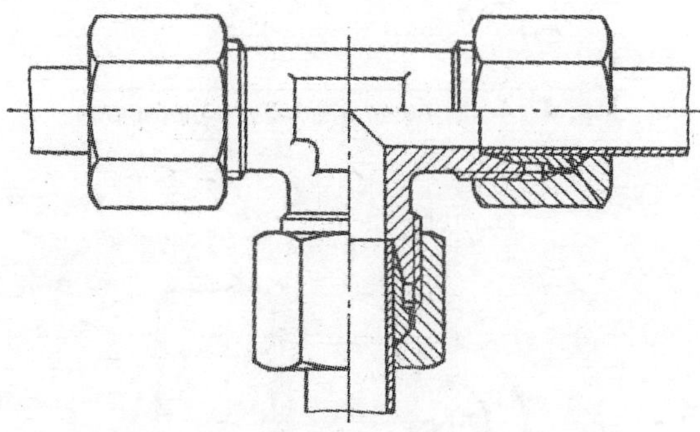

Fig. 4.14. *(c) Unequal pipe to pipe connector.*

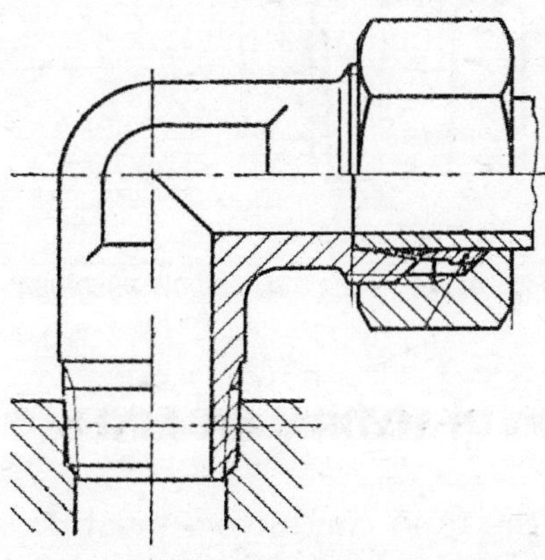

Fig. 4.14. *(d) Taper male stud 'T'.*

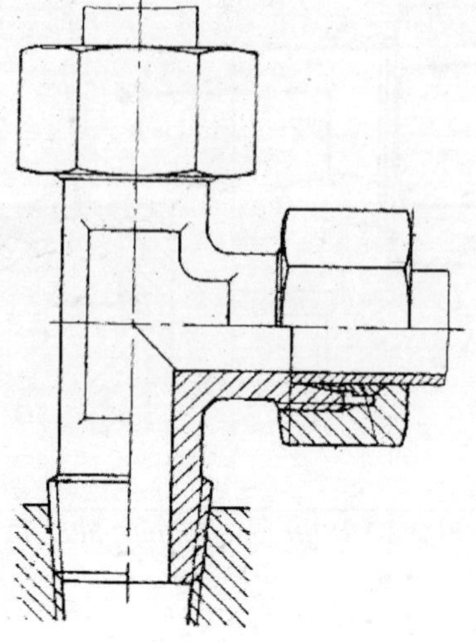

Fig. 4.14. *(e) Equal male union pipe-pipe.*

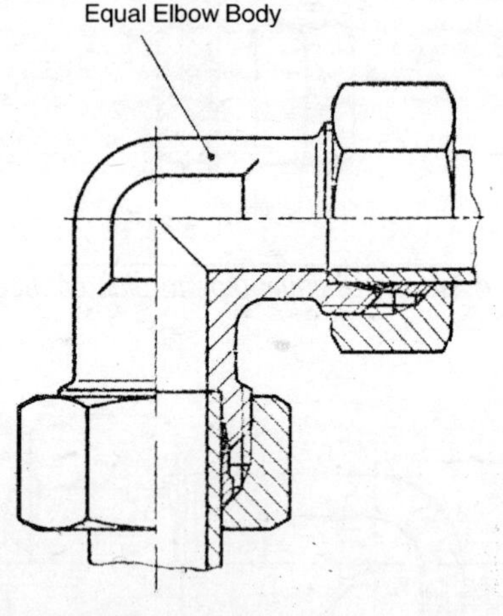

Fig. 4.14. *(f) Male run elbow assembly.*

4.8 INSTALLATION OF HYDRAULIC LINES

The inside diameter of the fluid conductors must be selected based on the recommended flow velocities for suction, pressure and return lines. The flow velocities recommended for suction lines is between 1 to 1.5 metres per second. The recommended return line velocities are between 1.5 to 4 metres per second. The recommended pressure line velocities are between 4 to 8 metres per second. Adhering to these standards will create a healthy hydraulic system by preventing cavitations and excessive pressure drops leading to inefficiency and heat generation.

The pressure line, inlet lines and return lines each have their own functions. The pump inlet port is usually larger than the outlet port to permit free flow of oil by reducing excessive intake velocities for a particular discharge. While the length of the inlet line must be as short as possible, their diameters must be as large as specified. Number of fittings and bends used on the intake side must be minimum. Intake lines must be air tight, specially, in case of pumps mounted above the reservoir. The intake pipes, which usually have a taper pipe thread on them, must be tightened fully and sealed using M-seals or any other sealing compound. This would prevent completely air entering the system, improve volumetric efficiency of the pump and reduce pump noise considerably.

Unnecessary restrictions on the return lines will lead to build up of backpressure in the system resulting in heating up of oil and wasted horse power. Return line sizes must be adequate to allow low flow velocities and must terminate well below the oil-level in the reservoir. Otherwise the returning oil would splash creating aeration and consequent foam formation. The pressure side which consists of pipelines between the pump and the actuator must be as short as possible and must have recommended internal diameter consistent with the flow, so that recommended flow velocities can be maintained which will avoid excessive pressure drops, turbulent flow and heating up of oil. The threads on the pressure and return line fittings must be wound with Teflon tapes and must be tightened by placing copper or bonded washers. Bonded washers besides providing a leak proof seal also enable the fittings to be positioned at convenient angles, as they need not be tightened all the way through to prevent leak. This has a definite advantage when tightening elbow fittings, which needs to be positioned at a convenient angle.

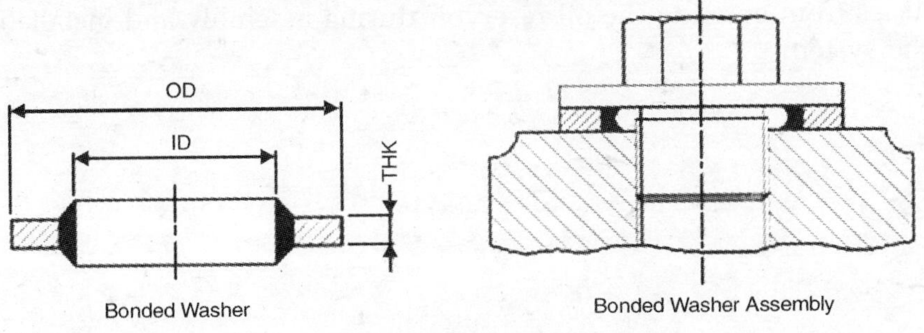

Fig. 4.15

4.9 CLEANLINESS

Contamination is perhaps the only cause of premature system malfunction and failure in a well conceived and designed hydraulic system. Apart from proper assembly and right installation, cleanliness is equally important for the reliability of the system. During assembly, it is necessary that the hydraulic parts be assembled in a clean environment using technician and workmen who have been trained. The technician and workmen involved should be made to appreciate the importance of cleanliness at every stage during assembly and installation of hydraulic system. The piping have to be pickled, neutralized and then washed with kerosene. The pickling will eliminate rust, scale and welding borax. The piping must be phosphate treated if necessary.

The oil reservoir is usually a fabricated structure therefore, it becomes very important to clean the oil reservoir completely of all the contaminating particles. Sand blasting is a sure and efficient method of removing rust and scale on hydraulic pipes and fittings. However it should be ensured that sand particles sticking to crevices and blind holes in the pipes and fittings are thoroughly flushed out. If the pipes were long and bent into complex shapes, then instead of sand blasting, pickling would be an ideal solution.

It must be ensured that all openings in to the hydraulic system are properly covered to keep out dirt and metal parts, when work such as drilling, tapping, welding is being carried out in the vicinity. Particles and micro-particles, which circulate continuously in the oil, are responsible for wear and tear of the reciprocating and rotating surfaces of the hydraulic components. So it is important to ensure that the hydraulic oil is free of these particles by filtering them at the time of filling up the tank. Hydraulic pumps are invariably fitted with strainers on the intake lines and filters on the pressure and return lines to prevent contaminated particles from entering the system. A separate chapter is devoted to this elsewhere in this book. Here the emphasis is on preventing contamination, dirt, abrasive particles from entering the oil reservoir during assembly and installation of the system.

5

POWER CONTROLS – VALVES

In oil hydraulic systems we have a wide variety of valves to perform different control functions such as, (1) open or close the passage of oil through a certain line, (2) control the direction of oil flow, (3) control the quantity of oil flow to and from the actuator, (4) control the pressure developed by the system, (5) un-load pump delivery to tank at low pressure when the system is idling, (6) sequence the flow of oil to two or more actuators. (7) Counter-balance a load against gravity etc.

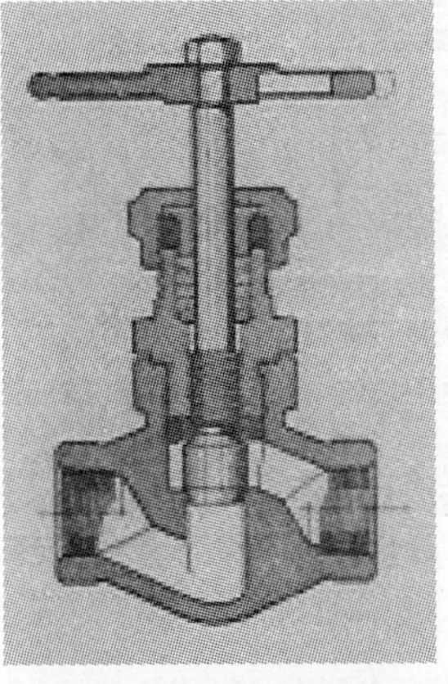

Fig. 5.1. *(a) Gate valve.*

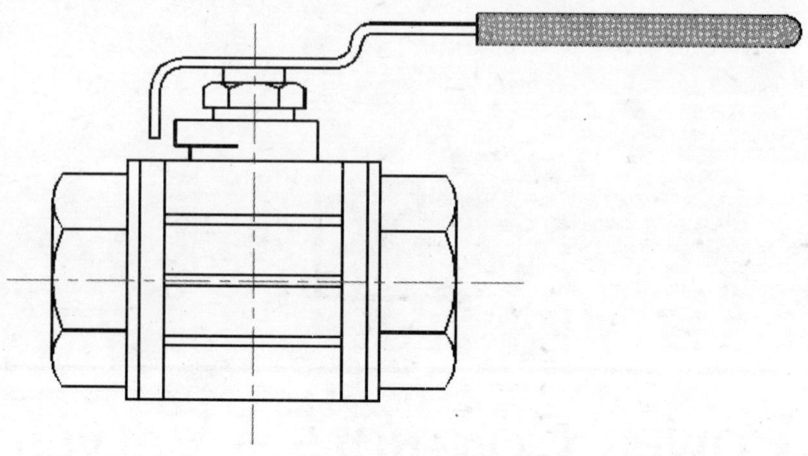

Fig. 5.1. *(b) Ball valve.*

The all too familiar *gate valve* or *ball valve* can keep an oil passage line completely closed, open or partially closed. The actuation is usually through a hand lever/hand-wheel or hand knob. They are used both in high-pressure and low-pressure lines. Low-pressure valves are usually gray iron or brass castings primarily used in suction lines to isolate an over-head oil tank to facilitate maintenance and servicing. If these valves are used in the suction line it is important to ensure every time, before starting the pump that the valves are fully open, to prevent damage to the pump. High-pressure valves are made of alloy steel castings or stainless steel and are used on the pressure/return lines.

Control of direction of oil flow is a vital function in a hydraulic system. A simple check valve permits oil flow in one direction and prevents return flow. This could therefore be called a directional control valve in the strict sense of the word.

A check valve in its simplest form consists of a steel body, which is usually hexagonal in shape with female pipe threads on either side to accommodate pipe or hose end connectors. A puppet, which opens or closes an oil passage, a compression spring and a spring retainer complete the assembly. The direction of oil flow is the direction in which the oil pressure can displace the puppet against the spring force to open the passage.

The pressure required to displace the puppet is called the cracking pressure. Check valves perform many other functions in addition to preventing return flow.

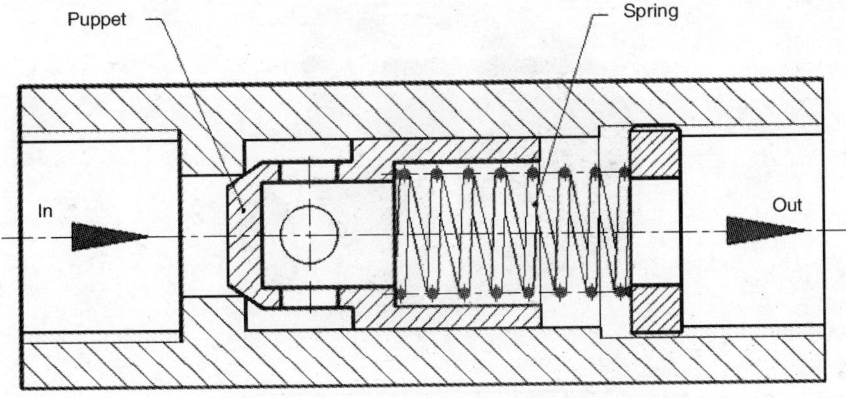

Fig. 5.2. *Cross section of in line check valve.*

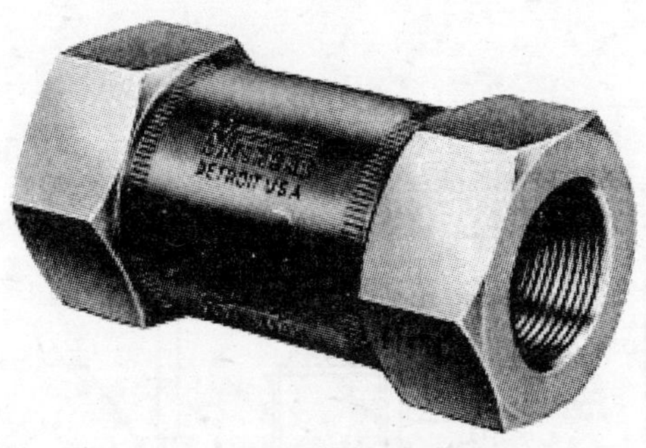

Fig. 5.3. *Inline check valve.*

Cracking pressure ratings differ accordingly. In check valves intended only for preventing return flow, the cracking pressure need be no more than 0.5 kgf/sq-cm or 0.5 bar. In check valves intended to create a pilot pressure that will actuate some other valve the cracking pressure required could be as high as 5 bar. The flow to be handled, the cracking pressure required and the maximum system pressure have to be specified to obtain the right check valve.

If the inlet and outlet passages are placed in the same line, it is called an *in line check valve.* If they are at right angles to each other, it is called a *right angle check valve.* A right angle check valve is used in a flow line where high velocity return flow is expected.

Fig. 5.4. *Right angle check valve.*

Typical Check valve applications where the right angle check valve ***must be used*** are given below. Check valves used in all the four applications shown in Fig. 5.5 are right angle check valves.

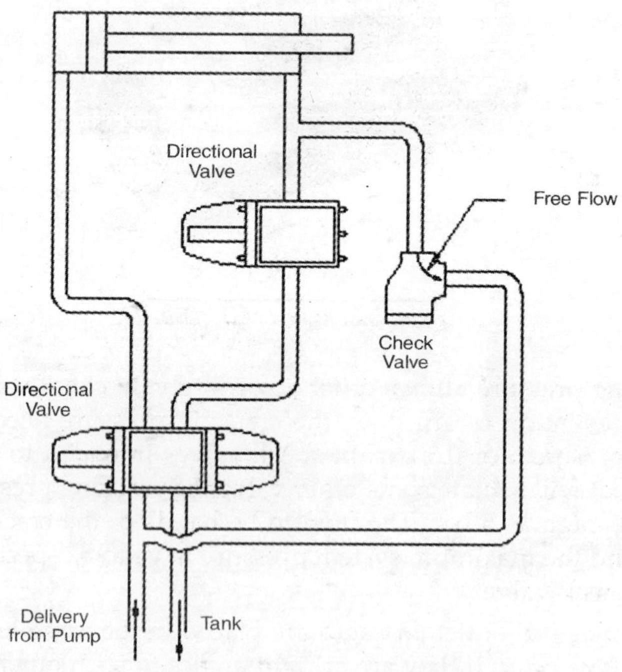

Fig. 5.5. *(a) Regenerative circuits.*

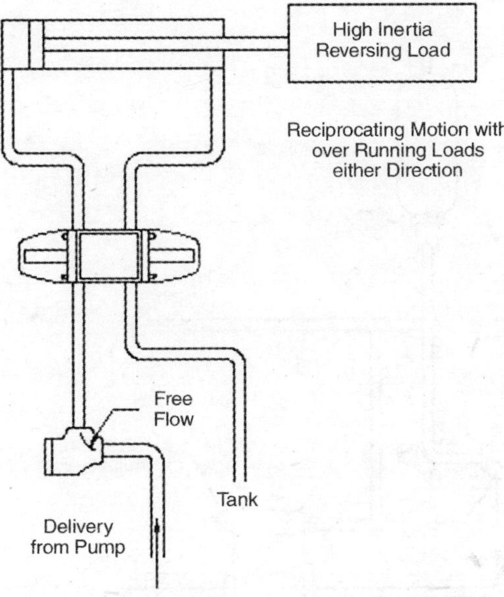

Fig. 5.5. *(b) High inertial loads.*

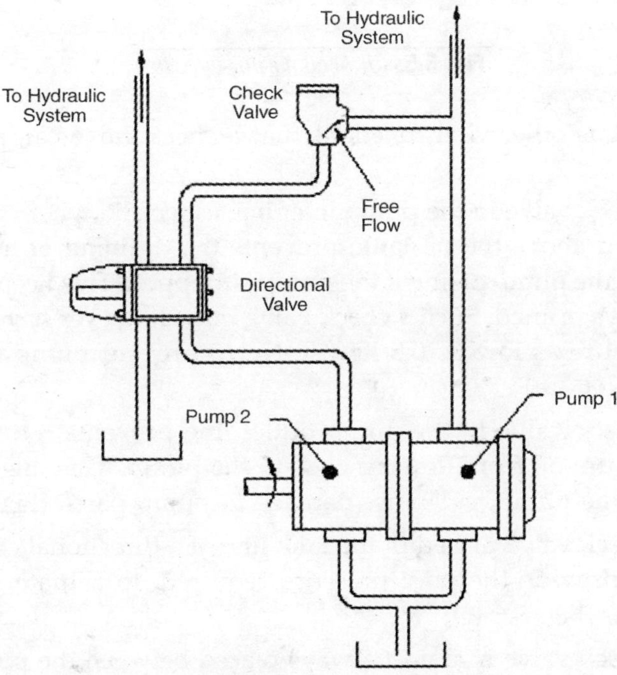

Fig. 5.5. *(c) High-low circuits.*

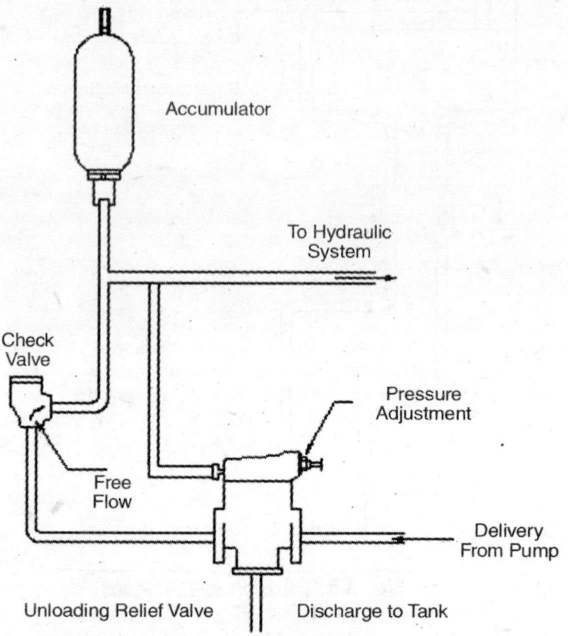

Fig. 5.5. *(d) Accumulator circuits.*

The various other vital functions that a check valve can perform are listed below.

- A check valve in the pump inlet line especially where the pump is placed above the oil tank, prevents the draining of hydraulic oil from the pump casing if the pump is stopped. This keeps the pump always primed. Such a check valve must however have a cracking pressure as low as 0.5 kgf/sq-cm to prevent pump cavitations. [Fig. 5.6 (a)]
- A check valve in the pump outlet line prevents returning high-pressure oil from flowing back to the pump. This might in effect, stall the pump and cause damage to pump parts. [Fig. 5.6 (b)]
- A check valve placed in the tank line of a directional control valve will provide the pilot-pressure required, to actuate some other valve. [Fig. 5.6 (c)]
- A check valve is almost always placed between the pump and an accumulator to prevent, high-pressure oil from flowing back to the pump. [Fig. 5.6 (d)]

- In dual pump systems employing high-pressure low delivery and low-pressure high delivery circuits, a check valve is used to isolate the high-pressure line from low-pressure line. [Fig. 5.6 (e)]
- Check valves in combination with flow control valves can offer a variety of economic solutions.
- A restriction check valve has an orifice drilled through the puppet to permit a restricted amount of return flow. This feature is used to achieve controlled de-compression in large hydraulic cylinders.

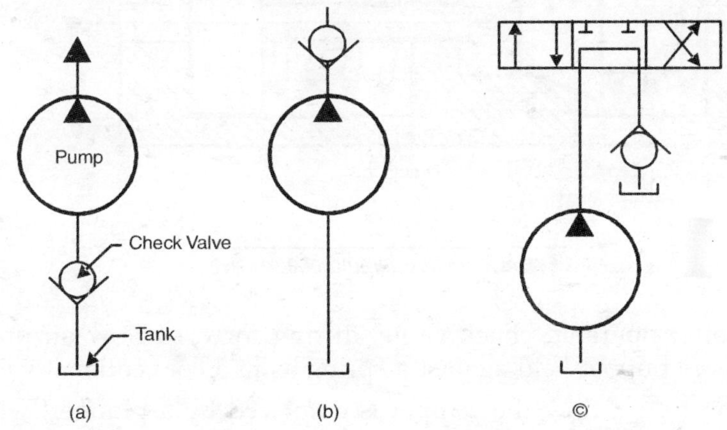

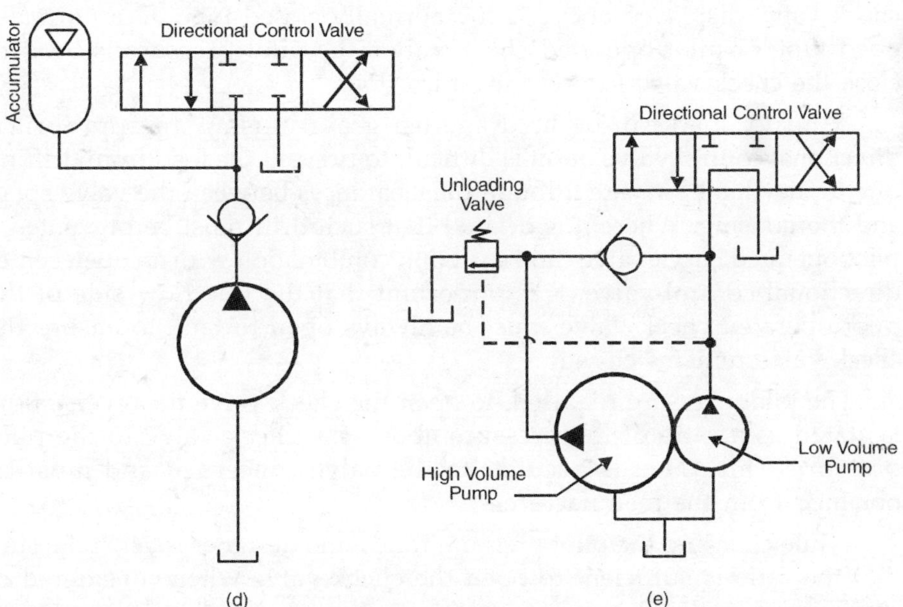

Fig. 5.6

There are check valves that permit return flow upon obtaining an oil pressure signal from a line different from the line in which the check valve is located. These are called *pilot pressure operated check valves.*

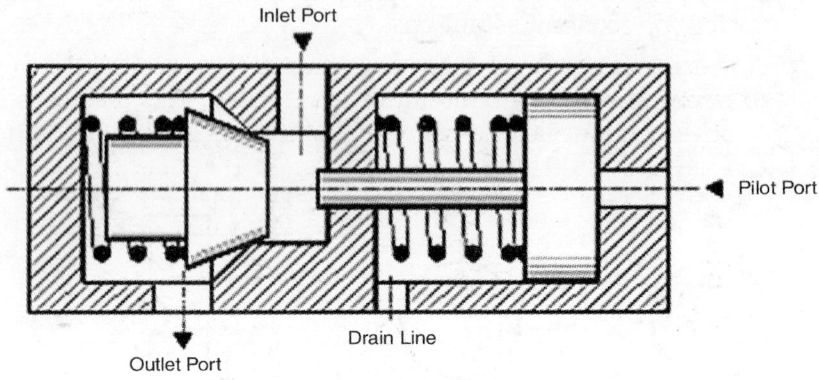

Fig. 5.7. *Pilot operated check valve.*

In a pilot operated check valve, during forward flow oil pressure displaces the puppet held against a spring as in a conventional valve.

During return flow the puppet is displaced by a plunger, which is actuated by pressure tapped from a secondary line. The pilot operated check valve displayed above is the normally closed type. In a normally open type of pilot operated check valve, the pilot pressure is used to close the check valve to prevent return flow.

A heavy load can be held against gravity using an appropriate directional control valve but it is difficult to prevent the load from drifting due to inevitable leakage through the clearances between the valve spool and the housing. Where it is desired that such drift must be prevented a pilot operated check valve can be used in combination with an open center directional control valve. It is important that the free flow side of the pilot operated check valve must be always open to tank to ensure the check valve remains closed.

The pilot pressure required, to open the check valve for reverse flow is stated as a ratio of the pressure above the check valve to the pilot pressure. This value is peculiar to the valve concerned and must be obtained from the manufacturer.

While choosing the pilot pressure ratios, the designer must make sure that the ratio is sufficient to open the check valve when considered in combination with the area ratios between the head end and the rod-end of the actuator. A pilot operated check valve, placed in the rod end side of the actuator must have a ratio larger than the cylinder area ratios for the check valve to open. The check valve ratio should not be too large either as this would create, back pressure in the tank return line. This can prevent the check valve from closing.

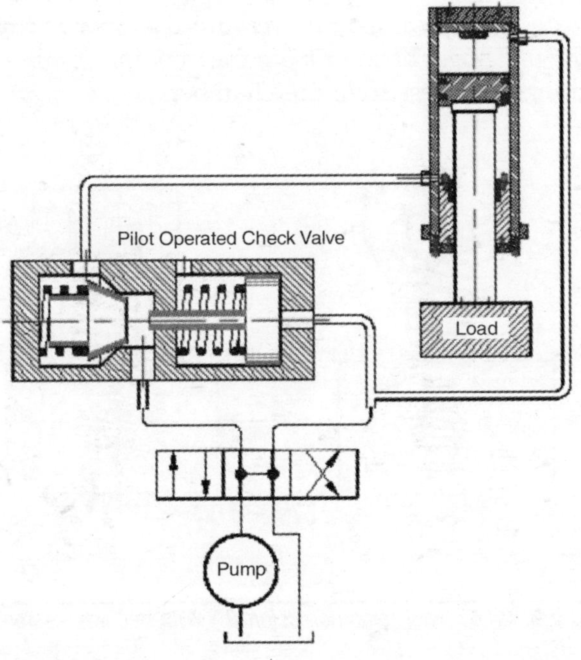

Fig. 5.8. *Pilot operated check valve being used to hold load against gravity.*

5.1 PRE-FILL VALVES

In a hydraulic system the speed of an actuator depends on the pump discharge. Higher the pump discharge, higher will be the speed for the same piston area. In large hydraulic presses the piston area can be so large that it requires huge deliveries even to maintain a decent speed.

Let us for instance consider a 30 cm diameter piston which is required to move @ 5 cm/sec. The pump delivery Q in liters/minute required to maintain this speed is, Q = [0 .7854 × (30) ^{2} × 5 × 60]/1000 = 212 LPM. Supposing the system operating pressure is 200 kgf/sq-cm, then the in put power required would be (200 × 212) / 600 = 70 kW without taking in to account the volumetric and the mechanical efficiencies of the pump. Such huge in put power is bound to affect the power factor and the connected load of the concerned unit besides resulting in wasted power. Even if we assume that in a cycle time of one minute the ram is pressurized for only 5 seconds, approximately, 45 kW-hour is consumed in a shift of

8 hours. In order to avoid such colossal waste of energy "pre-fill valves" have been designed. *A pre-fill valve* as the name suggests is a valve used to fill up the pressure chamber with oil directly from the tank through gravity or sometimes through vacuum pressure. This way the need for a large pump is eliminated. As soon as the pressure chamber is full, the pre-fill valve isolates the pressure chamber from the tank line. Only a small pump will now be enough to pressurize the chamber.

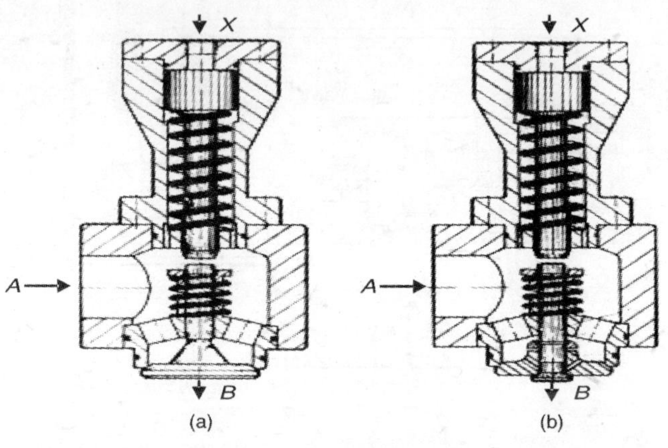

Fig. 5.9. *(a) Without decompression (b) With decompression.*

Pre-fill valves are invariably used in large hydraulic presses and are functionally identical to a pilot operated check valve. Since pre-fill valves handle large flows they are bigger in size compared to normal pilot operated check valves. A pre-fill valve may incorporate an additional feature to obtain controlled decompression of the hydraulic ram. A 'pre-fill' valve mounted on the cylinder will allow free flow of oil to the cylinder from tank port *A* to cylinder port *B* (See Fig. 5.10) through an overhead oil tank at the rate at which the ram could advance.

The speed at which the ram advances is now governed by the pump delivery directed to a small slave cylinder which is mechanically connected to the main ram As soon as the ram movement stops due to work resistance, the pilot pressure at port *X* that kept the valve open is withdrawn. This closes the 'pre-fill' valve thus isolating the tank from the cylinder. Only a small pump discharge, say about 10 LPM, is now enough to pressurize the same ram to 200 kgf/sq-cm. The result is, enormous savings in HP.

The decompression feature that enables the valve to open progressively in two stages for smooth and rapid exhaust is almost always desirable for large rams. Although this feature increases the response time of the valve it prevents large amount of high-pressure oil from being unloaded to tank every time the ram is decompressed.

Pre-fill valves can be bought out or manufactured 'in house'. The velocity of oil entering the cylinder during pre-filling should be ≤ 2 metres/second. The size of a pre-fill valve depends on the cylinder volume and the rate at which this volume must be filled. This is determined from the flow passage area available. A ram of 30 cm diameter has an area ≈ 700 cm^2. If this ram strokes for a length of 50 cm then the volume requirement would be 35,000 cc. If the stroke has to be completed in a time interval t = 3 seconds then rate of flow through the pre-fill valve would be 35,000/3 = 11,666 cubic centimeters per second. Since $Q = A \times V$, where A is the required flow passage area and V is the recommended flow velocity in cm. A = 11,666/200 = 58 cm^2 or ~ 86 mm diameter.

5.2 DIRECTIONAL CONTROL VALVES

The rotary or linear motion of an actuator cannot be instantly reversed or stopped or an actuator instantly depressurized without an appropriate directional control valve in place. Directional control valves therefore control the motion and the direction of motion of an actuator by diverting the pump flow to appropriate ports in the actuator.

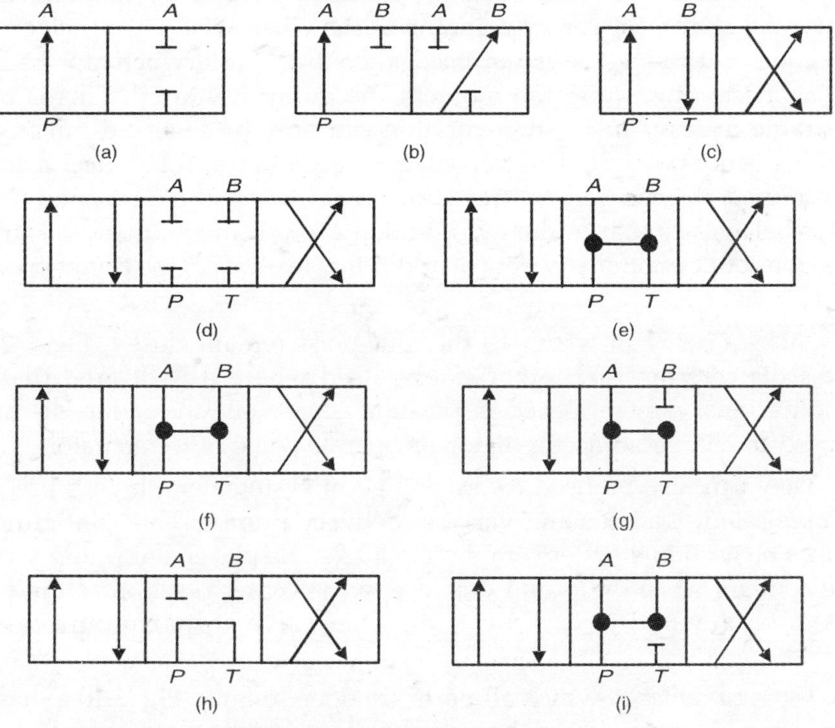

Fig. 5.10. *Valve configurations.*

Any directional control valve has a minimum of two positions, which it assumes alternately. A two-position valve may have 2, 3 or 4 pathways for the fluid. In the envelope for the valve shown (Fig. 5.10 a) flow from *P*→*A* is possible in one position. In the other position both *P* and *A* are blocked. This therefore is a two position-two way valve. In the envelope for the valve shown (Fig. 5.10 b) flow from *P*→*A* is possible in one position and in the other flow from *P*→*B* is possible. This therefore is a two position-three way valve. There is no pathway to tank here. Only a drain line to tank may be required to prevent fluid collection in valve cavities due to leakage past the spool clearances. If the valve is designed to drain internally even this line is not required.

A two position 4-way valve (Fig. 5.10 c) simultaneously connects *P*→*A* and *B*→*T* in one position and *P*→*B* and *A*→*T* in another position. This has a pathway, to tank available in either position. Here '*T*' represents the pathway to tank. *A* and *B* can be considered as pathways to the two sides of an actuator. So when the pump is serving the cap end of the cylinder, the rod end oil can exhaust to tank in one position and in the other, when the pump is serving the rod end, the cap end oil can exhaust to tank. This valve can be used to stroke a double acting linear actuator forward or return the actuator to its original position or can be used to rotate the shaft of a rotary actuator in the clock wise or anti-clockwise direction. Yet this valve is not used to control a rotary actuator because the shaft starts rotating the moment the pump is started. This is not a desirable feature. Also, shaft rotation can only be reversed but instant braking is not possible. This valve can be used to control a linear actuator if intermediate position control is not required. A three position 4-way valve behaves similar to the two position 4-way valve in the two extreme positions but can offer a variety of additional features in the center position of the valve.

A closed center in which all the four ports remain closed (Fig. 5.10 d) and isolated from each other can be used where it is desired that the actuator must remain locked in position even while the pump discharge is used to charge an accumulator or operate some other actuator.

They can also be used to advantage in systems employing pressure compensated, load sensing variable delivery pumps. This configuration cannot be used in a system employing a fixed displacement pump serving one actuator since the pump delivery in the center position of the valve has to be dumped to tank over the relief valve at maximum pressure setting.

The open center in which all ports are open to tank (Fig. 5.10 e) has the least pressure drop across the valve while dumping flow from Pump to tank. In the center position this valve cannot support the load against gravity. Also it is not possible to lock the actuator at desired locations. Open center valves can be used to control rotary actuators in either direction and linear actuators mounted horizontally if stopping the load midway is not an important criteria.

The configuration in which *P port is blocked while all other ports are open* (Fig. 5.10 f) cannot be used with a fixed displacement pump if the pump discharge cannot be used elsewhere during the center position of this valve. This valve again cannot support the load during the center position. In the center position the cylinders will float and the hydraulic motor will freewheel.

The configuration in which the ports *P, A and T are open and port B is blocked* (Fig. 5.10 g) can be used to control both the horizontal and vertically mounted cylinders. The valve can support a load against gravity in the center position and the piston can be braked at desired locations.

The configuration in which *P is open to tank while A and B remains closed,* (Fig. 5.10 h) also called the *tandem center* valve is by far, the most widely used configuration. Because the pump flow is connected to tank in the center position, pump can be idling at low pressure till the operator is ready to move the actuator. Because both *A* and *B* ports remain closed the load can be held against gravity and it is possible to inch the actuator to the desired position. Three different actuators can be controlled in sequence using three tandem-center directional control valves fed by one pump, which is not possible in any other configuration.

The pressure center configuration (Fig. 5.10 i) in which the tank line is blocked but the *P* line is connected to *A* and *B* is an ideal configuration for regenerative circuits. Many more configurations are available to suit a variety of applications but only a few basic configurations are considered here.

The flow pattern, between the center and the two extreme positions as the valve shifts is an important factor. Prior consideration to this factor will enable the designer to provide for transient pressure surges, which might otherwise damage the system. For instance the internal construction of a tandem center valve might be such that it blocks all ports between two positions (Fig. 5.11 a). A pressure relief valve installed between the pump and the directional control valve alone can protect such systems. It is also possible that the valve has an intermediate position which connects all ports to tank that is, open center cross over (Fig. 5.11 b). Such a valve although has no pressure surges, momentarily opens all ports to tank and therefore there could be pressure drop causing the load to drop. The flow pattern as the valve shifts is peculiar to the valve design and must be obtained from the manufacturer direct.

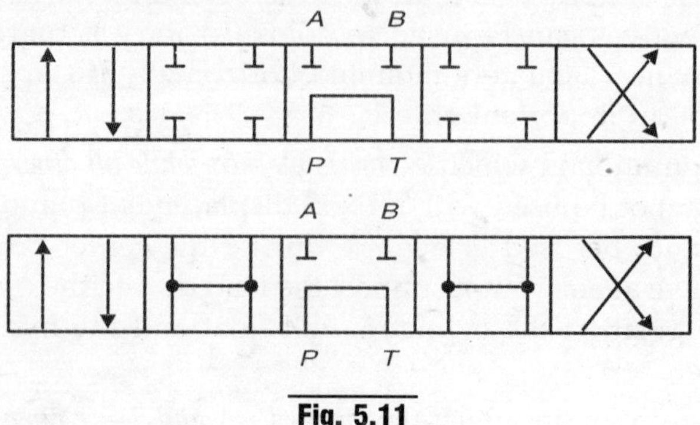

Fig. 5.11

Valves are either sliding spool valves or puppet valves or rotary valves. Most puppet valves are two position units. Puppet valves provide a leak proof closure. The puppet is in the form of a ball. In the starting position it is held on the valve seat by a spring thus keeping the passage closed. A plunger or a lever is used to displace the ball against the spring to open the passage. Puppet valves can provide a large flow area and therefore are suitable for large flows with minimum pressure drop. These however are not the most popular valves in use in the industry to day. A symbolic diagram of a two position, 3-way puppet valve in two different configurations is shown in Fig. 5.11 (c) and (d).

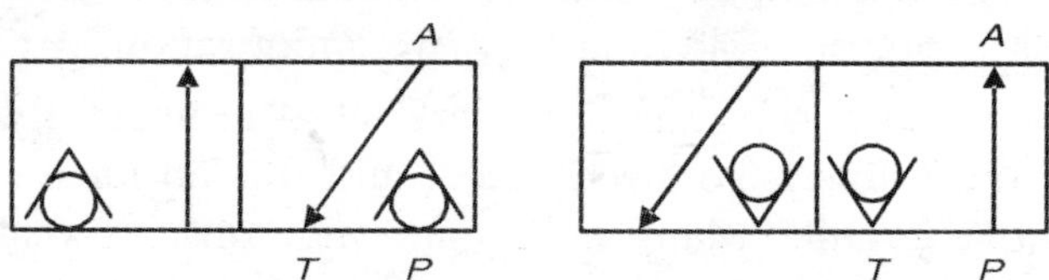

Fig. 5.11. *Puppet valves 2 position-3 way.*

Rotary valves, so called because it is the rotary motion of the spool that opens or closes a fluid passage, find extensive use in machine tool industry. They are good in applications where the flow is limited to 5–10 lpm.

Sliding spool valves are the most popular valves in use in the industry to day. Their biggest advantage is, one sliding spool can simultaneously open or close more than two passages unlike puppet valves that require a puppet/ball to control each passage. Because sliding spool valves depend upon close clearances between the spool and the bore to control leakage they are invariably not the preferred types in applications where the cylinder has to be held under pressure with no leak for a considerable time after cutting off the pump supply.

For such applications, special spool valves incorporating pressure-holding features are required.

In a sliding spool directional control valve a hardened and ground cylindrical steel spool slides in a machined bore in the valve body. The spool and the bore are matched and lapped together to maintain the minimum clearance required for the spool to slide in the bore. The valve body is usually a casting with cored passages. The lands in the spool are designed to cover/uncover a flow passage. The two way and four way versions of a sliding spool valve are shown in Figs 5.12 (a) and (b). The cast body is the same in either case. Only the land width of the spool used is smaller in case of the four-way valve to enable the spool to uncover the tank port in extreme positions.

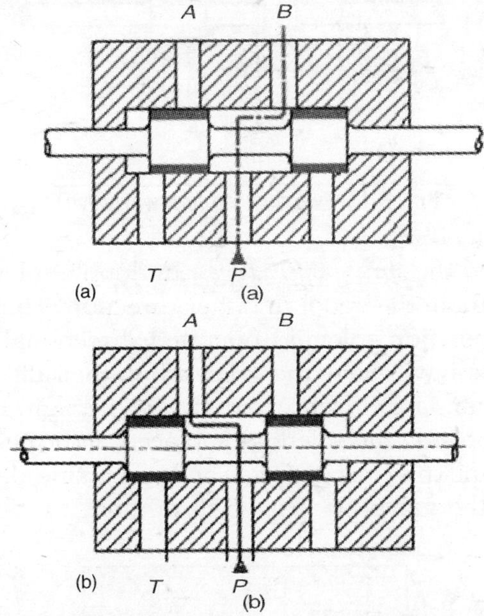

Fig. 5.12. *Two position two way valve.*

The shifting or actuation of spools in spool valves is carried out either manually using a hand or foot lever or mechanically with the help of a cam or electrically by energizing a solenoid that pulls the spool due to an electro-magnetic force. Accordingly they are designated as *hand/foot lever operated* valves, *cam operated* valves or *solenoid-operated* valves. Foot lever operated valves are usually rotary valves in which the spool instead of sliding, rotates in its cavity to open or close a passage. Rotary valves are usually used in machine tool applications. A two-position solenoid operated directional control valve might have single solenoid or two solenoids. Single solenoid versions depend on a spring to return the spool to its original position.

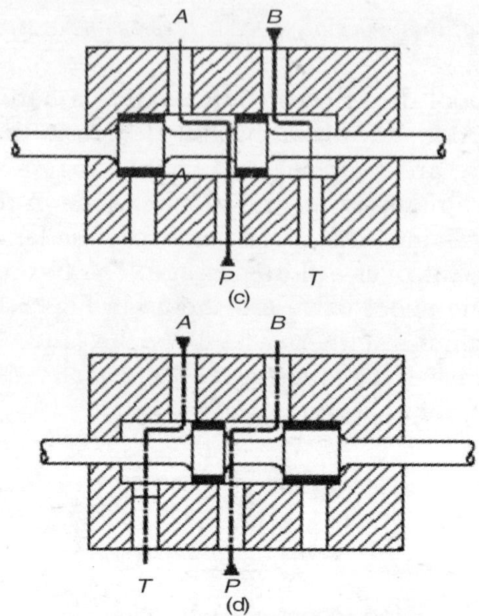

Fig. 5.12. *Two position four way valve.*

These are called the *spring-offset types.* In double solenoid versions the electromagnets actuate the spool in either direction. There are no springs involved. A two-position solenoid operated directional control valve of the spring-offset type will have the solenoid on the right side of the valve body. This is the standard offset. Alternatively, it might have the solenoid on the left side. The side the solenoid is placed, indicates the position the valve would assume after excitation. The spring side depicts the normal configuration for the valve. See Figs 5.13 (a) and (b).

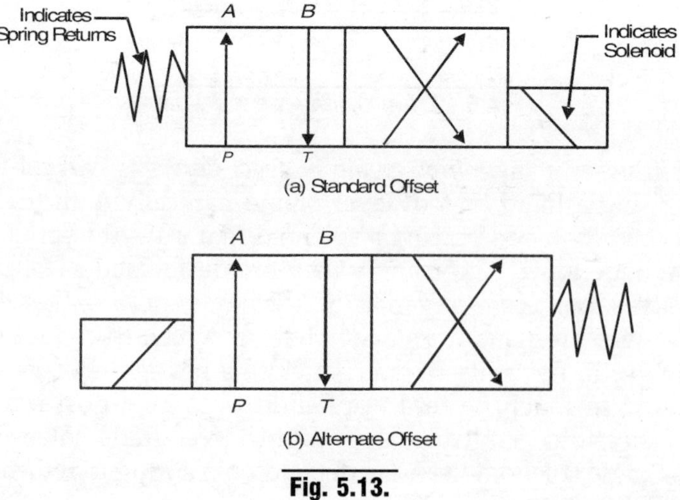

Fig. 5.13.

The three position four-way solenoid operated directional control valves comprise mainly the housing (1), one or two solenoids (2), control spool (3) and one or two return springs (4). Electro-magnetic solenoids (5) control the horizontal movement of the valve spool, which is basically a shift from one position to another to open or close a passage. These solenoids may be either air-gap solenoids or oil-immersed solenoids. The electro-magnetic force of the solenoids attracts control spool (3) by means of the plunger and shifts it from its resting position to the required end position upon obtaining an electrical signal. This permits free flow from *P* to *A* and *B* to *T* or *P* to *B* and *A* to *T*. When the solenoid (2) is de-energized control spool (3) is returned to its resting position by return springs (4). Control spools are accessible from either end for manual displacement while trouble-shooting or in emergencies.

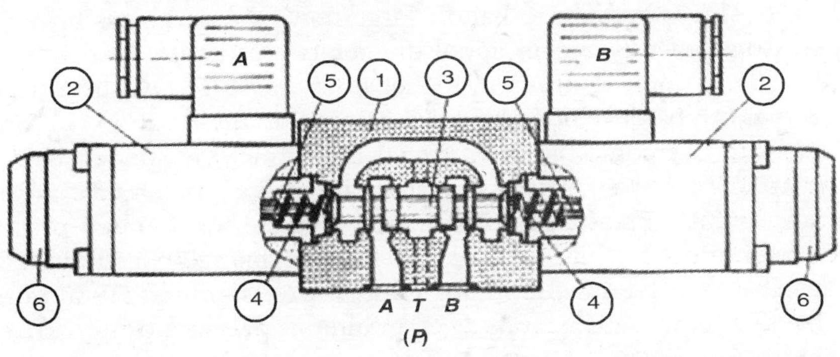

Fig. 5.14. *Three-position four-way solenoid operated directional control valve.*

Pressure tight oil-immersed solenoids also called 'wet armature solenoids' have an advantage over air-gap solenoids in that they are completely en-capsulated. The armature runs in oil, which results in low wear, good heat dissipation and a cushioned armature end stop. Both AC and DC solenoids are available. With the development of 'wet-armature' solenoids the problem of frequent solenoid burn out usually associated with 'air-gap' solenoids has largely been eliminated'. In situations where the springs fail to return the valve due to contamination or when both valve solenoids are simultaneously energized due to electrical faults the 'air-gap' solenoids used to fail instantly due to insufficient cooling. 'Wet-armature' solenoids employ the continuously circulating system oil to dissipate the heat and therefore the cooling rate is faster. Also 'wet-armature' solenoids make much less noise during shifting or return.

DC solenoids have gentle switching even at high switching frequencies. They hold the armature in any position with out danger to the spool and are not sensitive to voltage fluctuations and can be connected to an AC input by means of a rectifier.

AC solenoids permit simple electrical switching sequence and short switching times. No special provision for contact protection at the switch is required.

Solenoid are available for a wide variety of voltages, 110,220,440 V—50 Hz or 115, 230, 460 V 60 Hz. They are also available in 12/24 V dc versions.

Generally horizontal mounting is preferred for all solenoid valves as there would be no additional forces tending to shift the spools one way or the other. However two position, no spring versions with double solenoids must be mounted horizontally.

Direct solenoid-operated valves are not built for flows in excess of 100 LPM. This is because to handle large flows the port sizes have to be larger. Consequently bigger spool diameters are required to cover and uncover these ports resulting in higher flow forces. The electro-magnetic forces that can be developed by solenoids within these valves will not be sufficient to overcome the flow forces and attract the spools. The flow indicated is the maximum flow that the valve can handle without malfunction at a pressure approximately one half of its rated pressure. The pressure drop while passing this flow is generally around 20 kgf/sq.cm. Therefore the ideal flow for these valves is around 30–40 LPM to be able to operate satisfactorily with minimum pressure drop and at the maximum recommended pressure. To handle large flows *solenoid controlled pilot operated valves,* have been designed. These are two stage directional control valves comprising of the pilot stage and the main stage. The pilot stage valve is a small solenoid operated directional valve secured to the main stage by screws.

Energizing the solenoids of the pilot stage valve will shift pilot stage spool to permit oil supply to either end of the main stage spool.

In case of spool configurations, where the pressure line is open to tank in the valve center position the pilot pressure required to shift the main stage spool is obtained by placing a 5 bar check valve in the tank return line of the valve in case of internal pilot and in case of external pilot the check valve is placed in the pressure line between the valve and the pump and a tapping from here is taken to the valve sub-plate as shown in Fig. 5.15 (a) and (b).

The pressure exerted by this pilot oil will cause the main stage spool to shift thus influencing the direction of oil flow. These valves can handle very large flows up to 300 LPM and are available in as many configurations as the pilot stage valve.

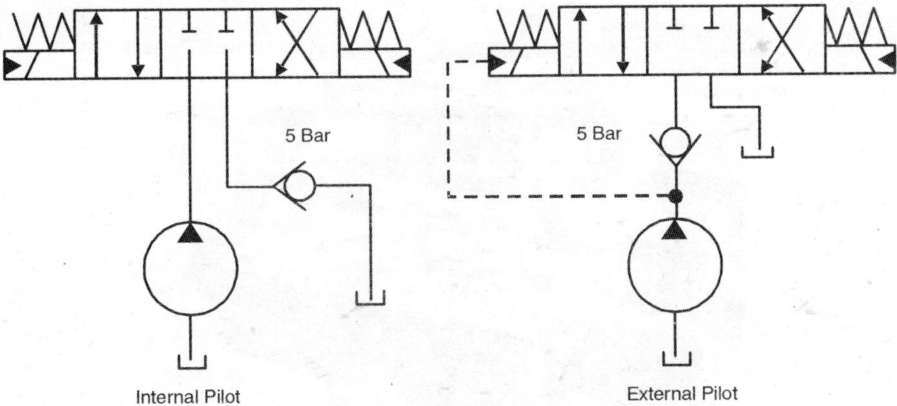

Fig. 5.15

Solenoid controlled pilot operated valves are usually equipped with controls to regulate the speed of spool shift from one position to another immediately following the pilot signal. This is called pilot choke control. A needle valve positioned in the pilot line controls the quantity of oil flow to large spool and thus its speed. This reduces shock and vibration. If pilot pressure can be tapped from a primary circuit *direct pilot operated valves* may be used to control the direction of oil flow in the secondary circuit thus eliminating the need for a pilot stage valve.

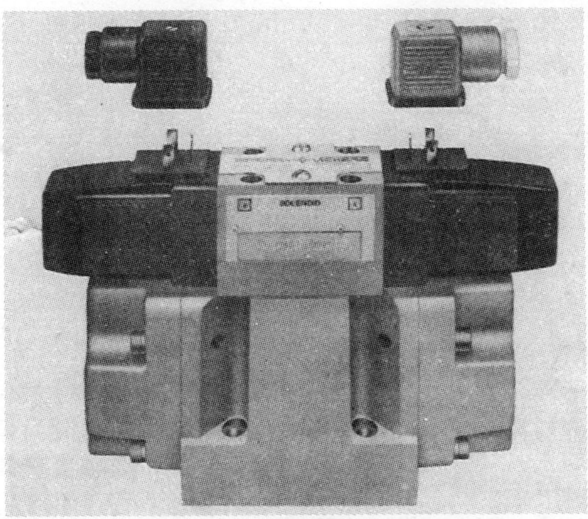

Fig. 5.16. *Solenoid controlled pilot operated four way three position directional control valve.*

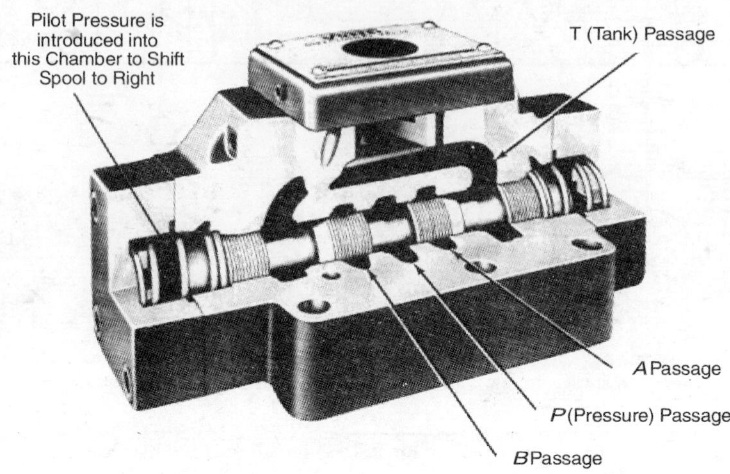

Fig. 5.17. *Pilot operated directional control valves.*

Hand lever operated valves are available as two-position/three-position valves either in *spring-returned* or *spring-detented* versions, in sizes to handle flows, as small as 10 LPM to as high as 500 LPM. In spring returned versions the hand lever must be kept depressed or pulled for as long as it is required. Taking the hands away from the lever would instantly return the spools to their original position. In spring detented versions the spool will stay put, in the position last attained until shifted manually.

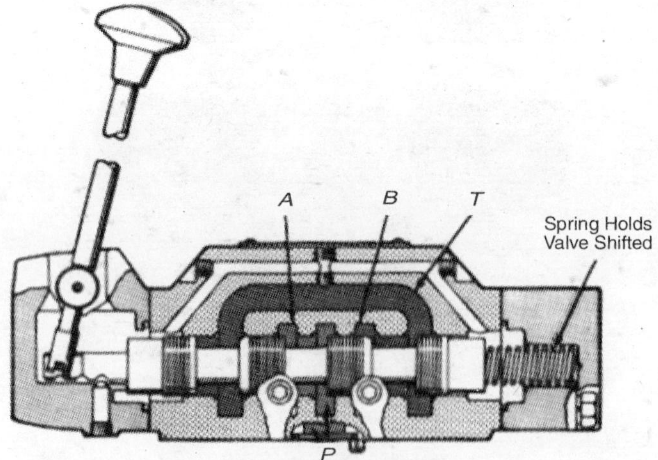

Fig. 5.18. *Manually operated valve.*

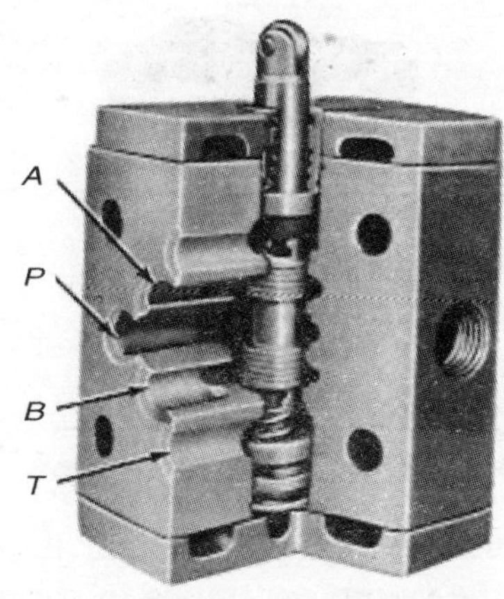

Fig. 5.19. *Mechanically operated valve.*

The mechanical *cam operated valves* are usually two-position spring offset type. The spool is shifted against the spring by a cam or any other mechanism. These valves can be used to automatically actuate a secondary system after the primary cylinder has completed its stroke. Directional control valves are either mounted on a sub-plate or on a manifold. The manufacturer of the valve will also supply the sub-plate to suit. Manifold blocks facilitate mounting of all components of the circuit on one solid block with internal connecting passages thus eliminating the need for a separate pipe connection for each valve.

The size of a valve is determined by the flow that it can handle with minimum pressure drop. *Valve pressure drop characteristic* therefore is an important feature that needs to be examined before selecting the directional control valves. This would reveal the back-pressure generated by the valve while passing the required flow. The maximum limit for this pressure drop is between 5–10 kgf/sq-cm when passing the system flow at the designed pressure. This will improve overall system efficiency, reduce heating up of oil and increase the life of the solenoid. Valve pressure drop characteristics can be obtained from the manufacturer.

5.3 FLOW CONTROL VALVES

Pump displacement or volume flow rate expressed in cubic centimeters/second controls the linear or rotational speed of an actuator. Volume flow rate in a fixed displacement pump is constant regardless of the requirements of the actuator.

Fig. 5.20. *Flow control valve.*

Even after the piston reaches the 'dead end' beyond which no further movement is possible the pump continues to supply the same quantity of fluid thus necessitating the need for a relief valve to dump the excess oil to tank. Volume flow rate in a variable displacement pump is however need-based. The internal mechanism of the pump regulates itself to meet the actuator demands at any point in the work cycle. Although they can regulate the flow to an actuator, they respond only to load resistance and therefore do not offer solutions to every circuit requirements. Flow control valves have therefore become necessary to control the speed of an actuator especially in circuits using fixed displacement pumps. The speed control is achieved by placing a flow restrictor in the fluid line to the actuator, between the actuator and the pump. The point to appreciate here is, placement of a flow restrictor as such will not cause reduced flow to the actuator because at any two points flow remains constant according to Bernoulli's principle $a_1v_1=a_2v_2$ where a and v represents area and fluid velocities upstream and downstream of the restrictor. The reduction in flow is caused by the pressure energy in the flow line upstream of the restrictor that opens the system relief valve through which excess oil is diverted to tank. The flow restrictor could be a simple orifice, such as a drilled hole in a fitting in which case it becomes a fixed orifice or it can be a needle valve whose flow passage area could be varied. By varying the passage area it is possible to regulate the flow to the actuator within the limits of the cracking pressure and the full flow pressure rating of the relief valve. If the amount of fluid flowing into the cylinder is controlled then it becomes the ***meter-in*** flow control. If the amount of fluid flowing out of the cylinder is controlled then it is called ***meter-out*** flow control. Meter-in flow control can be achieved by installing the flow control valve in the line between the pump and the directional control valve.

Meter-out flow control can be achieved by installing the flow control valve in the tank line of the direction control valve. Such a placement would ensure that the speed control is achieved both during 'approach' and 'return' strokes. However, if flow control is required in only one direction then the valve should be installed in the line between the actuator and the directional valve. Where flow control is required in only one direction then a check valve must be placed in parallel to the flow control valve to permit free return flow. Flow control valves with integral check valves are also available.

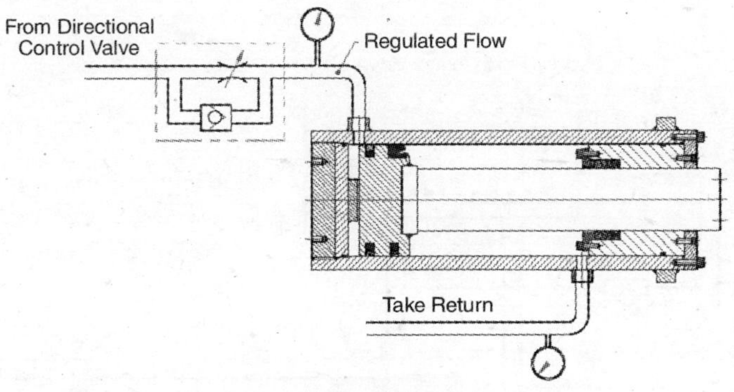

Fig. 5.21. *Meter-in.*

Meter-in flow control is recommended for actuator speed control in an open center circuits. Meter-out flow control is recommended for all applications that create negative or over-running loads such as when lowering heavy weights. This is because the flow resistance is in a direction opposed to the load movement. There is no difference in the mechanism of flow control achieved in either method. The difference lies only in where the valve is placed in the circuit to achieve flow control.

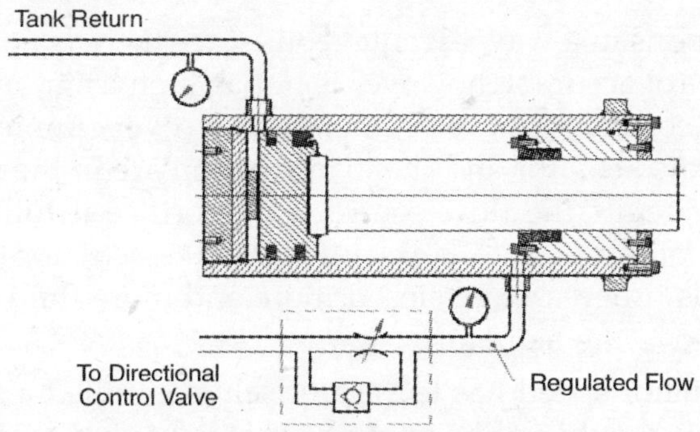

Fig. 5.22. *Meter-out.*

Bleed-off is another method of controlling the flow to the actuator. In this method the excess oil is by passed to tank directly by the flow control valve itself instead of blowing it over a relief valve at system pressure. There is a difference in the mechanism of flow control achieved in this method.

The flow control valve used in this case is called the 'by-pass' flow regulator. The mechanism of action of a by-pass flow regulator is illustrated below.

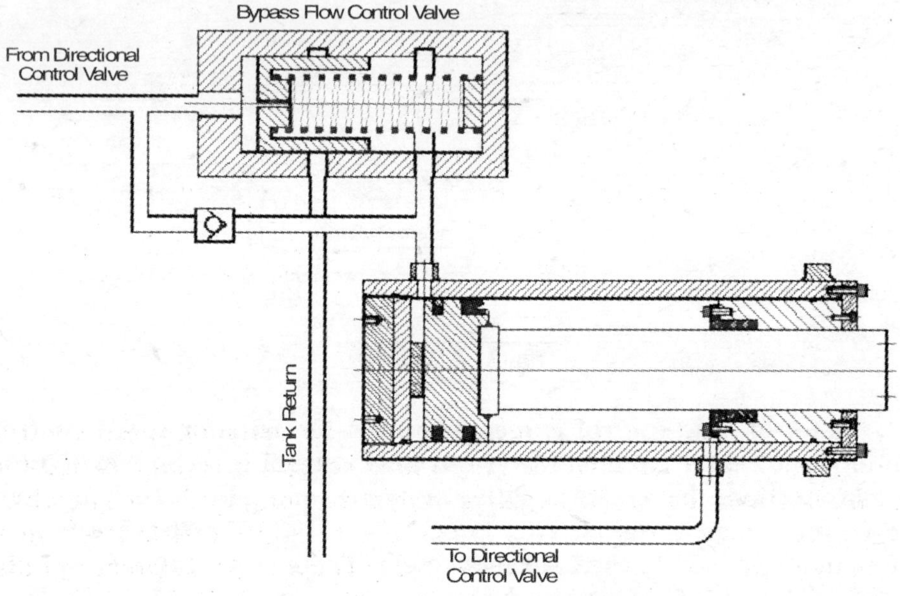

Fig. 5.23. *Bleed-off.*

This method is in a way advantageous since there is no wasted energy involved. Control accuracy however is not as high as in 'meter-in' circuit. Bleed-off flow control is not recommended for 'over-running' loads where the motion of the actuator and the load reaction are in the same direction. In 'bleed-off' circuits the flow control valve orifice if fully open will by pass all of the pump flow to tank and hence the actuator does not move. As the orifice is progressively closed more and more fluid would become available to move the actuator.

Where actuator speed has to be precisely controlled a simple variable orifice regulator would not be enough. This is because any change in the pressure drop across the valve due to load resistance would affect the feed rate since flow through an orifice is proportional to the square of the pressure drop. This lead to the development of the *pressure compensated flow control valves*. In this, there are two types. Fixed orifice pressure compensated, also called restrictive flow regulator and variable orifice pressure compensated flow control valve. In fixed orifice flow control valve the pressure drop causing the flow through the metering orifice is applied to either end of a pressure balanced compensator spool. Any variation in flow causes a variation in pressure drop. The resulting force imbalance moves the compensator to further restrict the available flow passage area thus throttling the flow to maintain the desired flow pattern.

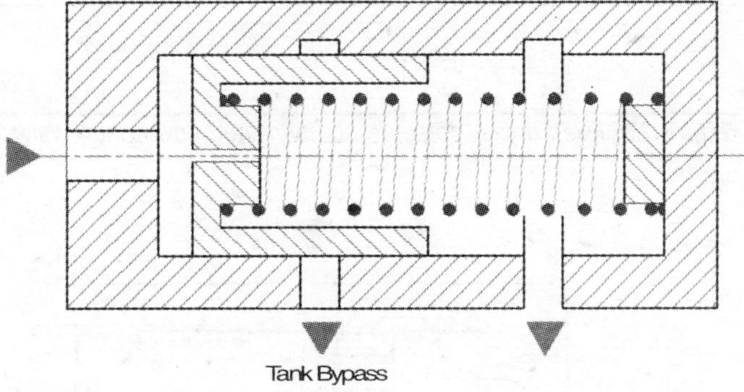

Fig. 5.24. *By pass flow regulators used in bleed-off circuits.*

The pressure compensated variable orifice flow control valve consists of a compensator spool biased against a spring that automatically adjusts itself to variations in pressure and a flow control spool to regulate the flow. Flow is regulated externally with the help of a set- screw that moves the flow control spool forward or backward so as to increase or decrease the flow passage area. Any tendency for the flow to vary beyond the set limit upsets the existing pressure differential that is keeping the compensator spool in position. The compensator spool moves to take up a new position. This will regulate the incoming flow to the flow control spool so as to maintain constant flow. Accuracies as high as ± 0.5% can be achieved with pressure compensated flow control valves. They are available with integral check valve to permit free reverse flow. Restricting the flow passage area of high-energy fluid is an inefficient but an inevitable method of speed control of hydraulic actuators. The consequent energy loss is converted in to heat raising the circulating oil temperature, which reduces the viscosity of the oil. This in turn affects the performance of the flow control valve. Obviously pressure compensation alone may not be enough if very accurate results are desired. Temperature compensation too, must be provided.

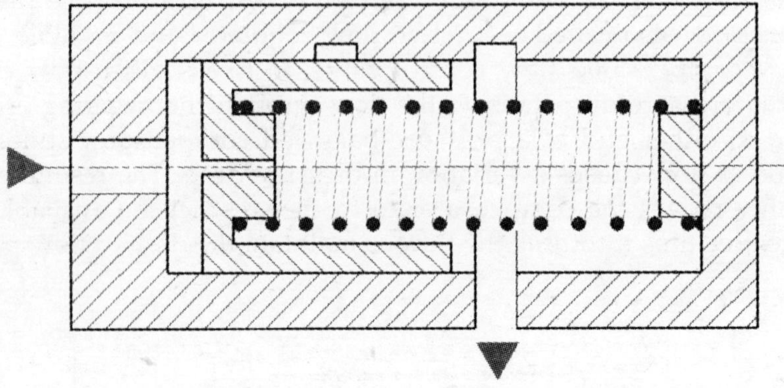

Fig. 5.25. *Fixed orifice pressure compensated flow control valve.*

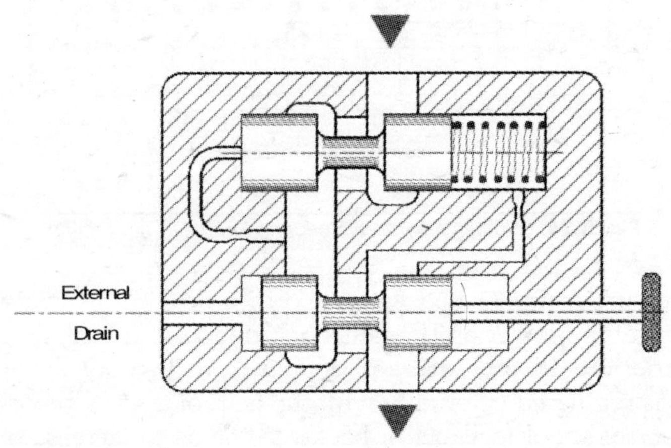

Fig. 5.26. *Variable orifice pressure compensated flow control valve.*

Combined pressure and temperature compensated flow control valves have, an additional compensating rod, which elongates in response to increased temperatures. This elongation closes the throttle opening and thus maintains constant flow across the valve. These valves are always provided with integral check valve for reverse free flow. Flow control accuracies as high as ± 2% is possible with these valves. They cost more and therefore their use is justified where such accurate controls are required.

A flow control valve must be large enough to handle system flows at minimum pressure drop. Therefore their maximum recommended controlled flow with good regulation must be ascertained. Also the pressure differential at which they can pass the minimum controlled flow is an important characteristic.

Their pressure rating must be consistent with the system requirements. All of these valves can be used in 'meter-in', 'meter-out' or bleed-off circuits.

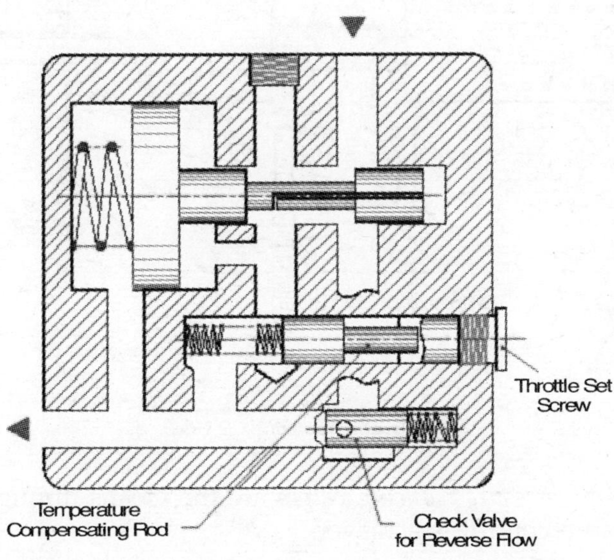

Fig. 5.27. *Pressure and temperature compensated flow control valve.*

The *deceleration valve* is one of the many variations to flow control valves. The function of the valve is to gradually decrease the actuator velocity towards the end of its stroke. It has a spring offset spool which has a cam roller fixed to it at one end. The cam attached to the piston rod gradually depresses the cam-roller attached to the spool to decrease the flow passage area. As the flow passage area decreases, the cylinder backpressure increases forcing the piston to decelerate.

5.4 PRESSURE CONTROLS

The pressure rating for any pump is the maximum value set by the manufacturer for continuous operation of the pump over its lifetime at the efficiencies indicated. Pumps especially piston pumps can exceed their pressure ratings till the prime mover stalls or some thing breaks with serious repercussions to both the machine and the persons nearby. Pressure control therefore is a basic requirement of any hydraulic system. *Pressure*

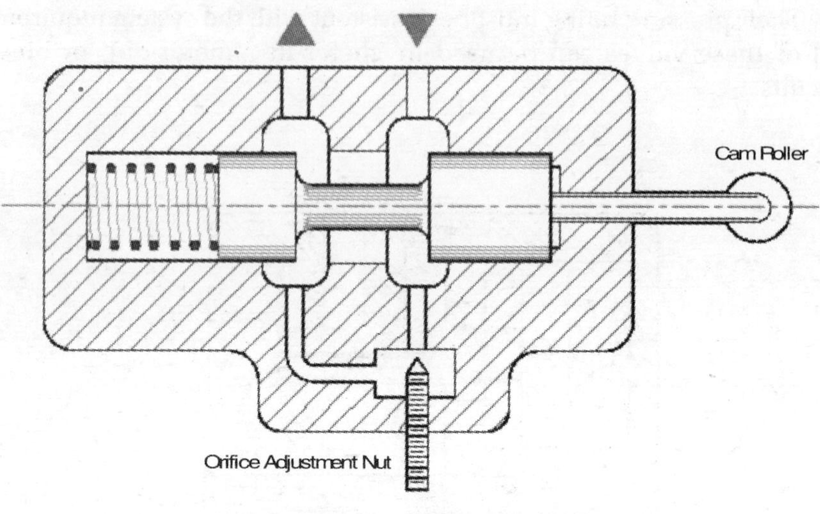

Fig. 5.28. *Deceleration valve.*

relief valves and *pressure reducing valves* are the means through which the system pressure control is achieved.

While a pressure relief valve is basic to any hydraulic system and cannot be dispensed with, the need for a pressure-reducing valve is based on the particular circuit requirement. A pressure relief valve controls the overall system pressure. Its function is similar to that of a fuse in the electrical system. A pressure-reducing valve serves a branch line that requires much lower pressure than the maximum system working pressure. Its function is more like the step down transformer.

A check valve with a stiffer spring can function like a pressure relief valve. When the pressure exceeds the spring rating the puppet is displaced from its seat allowing excess oil to 'blow over' to tank and thus limiting the system pressure. The only drawback with this arrangement is that infinite variation of pressure to any desired value is not possible.

If the check valve could be modified to make this possible then it becomes the *direct operated pressure relief valve* so called because the pressure acts directly on the puppet that opens to divert the flow to tank.

The additional feature required is the pressure adjusting mechanism, which is nothing more than a setscrew biased against the spring to vary the spring force. Some valves may be provided with a hand knob on the setscrew for convenience. Obviously the tank line is now placed at right angles to the axis of the puppet. This arrangement permits the pressure to be set to any desired value by tightening or loosening the setscrew. The setscrew is always provided with a lock nut to prevent inadvertent variation of pressure once the pressure is set. Within the limits of the spring stiffness the pressure can be set to any value.

The pressure at which the puppet first begins to divert the flow is called the 'cracking pressure.' As the pressure increases the puppet is further displaced from its seat permitting gradually, passage of most of the oil to tank. The difference between the full flow pressure and the cracking pressure is called the *pressure override.* Due to problems associated with large pressure override, direct operated relief valves are not recommended where a strict control over 'pressure-flow' characteristic is desired. They are also not used where large flows and high pressures are involved and where frequent pressure adjustment is required.

Balanced piston relief valve is technically superior to the simple direct operated relief valve. This is because the pressure override is much less and therefore th^ wasted horsepower.

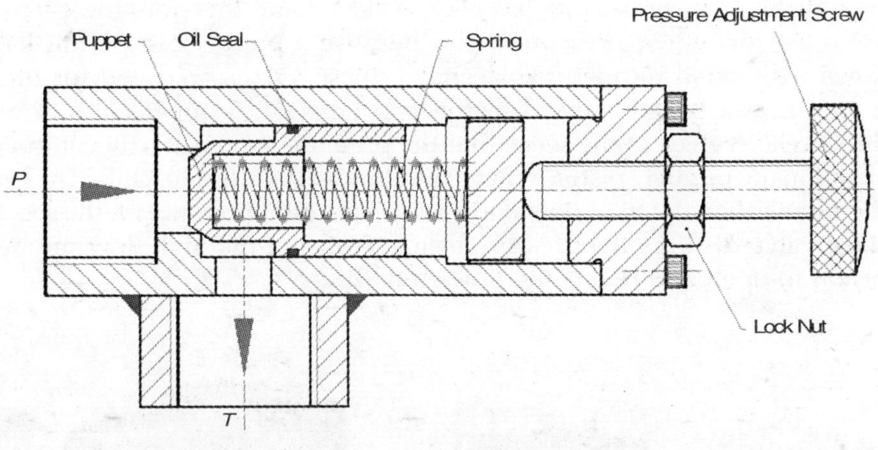

Fig. 5.29. *Direct operated pressure relief valve.*

Balanced piston relief valve is a two-stage valve. The first stage consists of a piston which has an orifice drilled through. This permits the piston to be in hydraulic balance because it is subjected to the same pressure on either side. This pressure is also sensed by the second stage valve, which is mounted on top. This is a simple direct operated relief valve of the type already discussed which has its tank line connected to the main stage valve through the balanced piston. A light spring enables the balanced piston to keep the main tank line normally closed.

When the pressure set in the second stage relief valve is reached the puppet opens reducing the pressure in the upper chamber by diverting the flow to tank through the cored hole in the balanced piston. The balanced piston is no longer in pressure balance. When the pressure differential between the upper and lower chamber reaches to about 1.5 kgf/sq-cm the balanced piston is lifted off its seat to divert the flow to tank. Since the valve opens to permit full flow at just about 1.5 kgf/sq-cm the pressure override is considerably reduced. These valves are ideal for handling large flows at medium pressures in the range of 175–200 kgf/sq-cm. Design engineers might prefer this valve for its obvious advantages but it is a maintenance engineer's nightmare due to its frequent failures for a very simple reason. The balanced piston orifice gets clogged with contaminant frequently. Therefore a filtration of 20 microns or less is a must while using these valves.

A variation to this valve is the *balanced puppet relief valve.* It is also a two-stage valve similar to the balanced piston relief valve with a low-flow adjustable cone section mounted on top of a high-flow pilot operated puppet section. Using a puppet instead of a piston that acts more like a spool valve, a better sealing has been achieved and therefore these valves are rated for higher pressures. Also because a puppet can permit large flows with small puppet displacement, these valves are rated for much higher flows. Both balanced piston and balanced puppet vales are in-direct relief valves in the sense that the pressure is not directly controlled at the main puppet/piston where the flow is diverted to tank. They are therefore called the pilot operated valves. The pilot pressure on the second stage valve disturbs the pressure balance causing the high flow puppet/piston to open and divert the flow to tank.

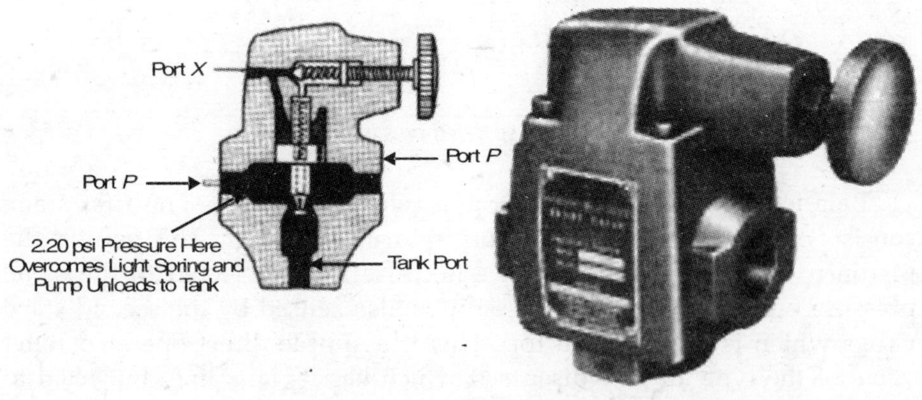

Fig. 5.30. *Balanced piston relief valve.*

Flow from port *A* to tank port *T* remains blocked as long as the high flow puppet is in pressure balance due to the oil pressure being sensed on either side of the puppet through the orifice. The same pressure is also sensed at the cone of the adjustable relief valve in the upper chamber. So long as this pressure does not exceed the setting of the relief valve spring the large flow puppet remains in pressure balance and port *A* remains blocked. As soon as pressure at the cone of the adjustable relief valve exceeds the setting of the relief valve spring the cone opens permitting a little oil to drain to the tank thus reducing the pressure in the upper chamber. This upsets the pressure balance and the main puppet lifts passing only enough flow from *A* to *B* to maintain the pressure at the set value. Infinite pressure adjustment from 0–350 kgf/sq-cm is possible with the use of these valves. In these valves only a small quantity of fluid initially blows over at high pressure, the rest of the fluid is diverted to tank from *A* to *B* at a very small pressure differential. This causes appreciable reduction in wasted horsepower resulting in reduced heating up of oil. Pilot operated relief valves costs more than the direct operated relief valves and strict control over backpressure in the tank line must be maintained for the valves to function satisfactorily.

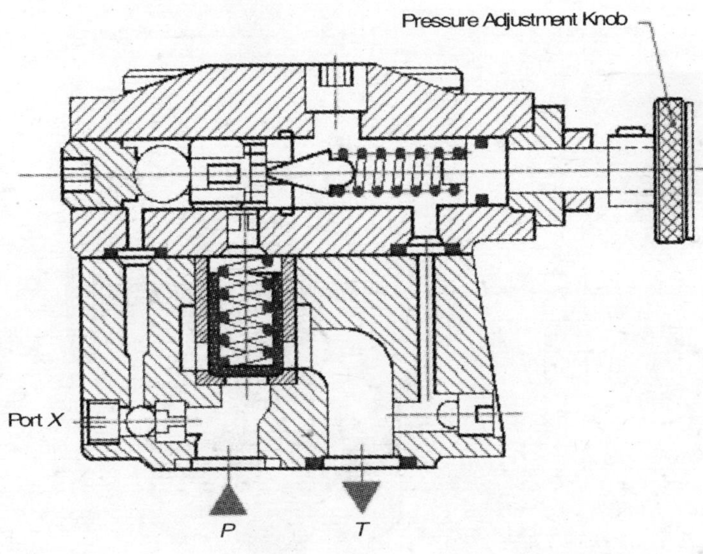

Fig. 5.31. *Balanced puppet pilot operated relief valve.*

Direct operated relief valves have only two ports, the pressure port and the tank port. Pilot operated relief valves have an additional port *X* which can be used to control these valves from a remote location in combination with a small direct operated relief valve. (Fig. 5.32 a). Both the valves of-course must have the same spring rating. In combination with a solenoid valve the port X can be used to vent these relief valves to tank at very low pressures even while providing an electrically controlled adjustable pressure relief. In the configuration shown in Fig. 5.32 (b) below the relief valve is normally vented to tank at low pressure. Only when the connection P to A is established, that is when the solenoid is energized the valve functions like a relief valve. By using appropriate solenoid valves a variety of control functions can be established.

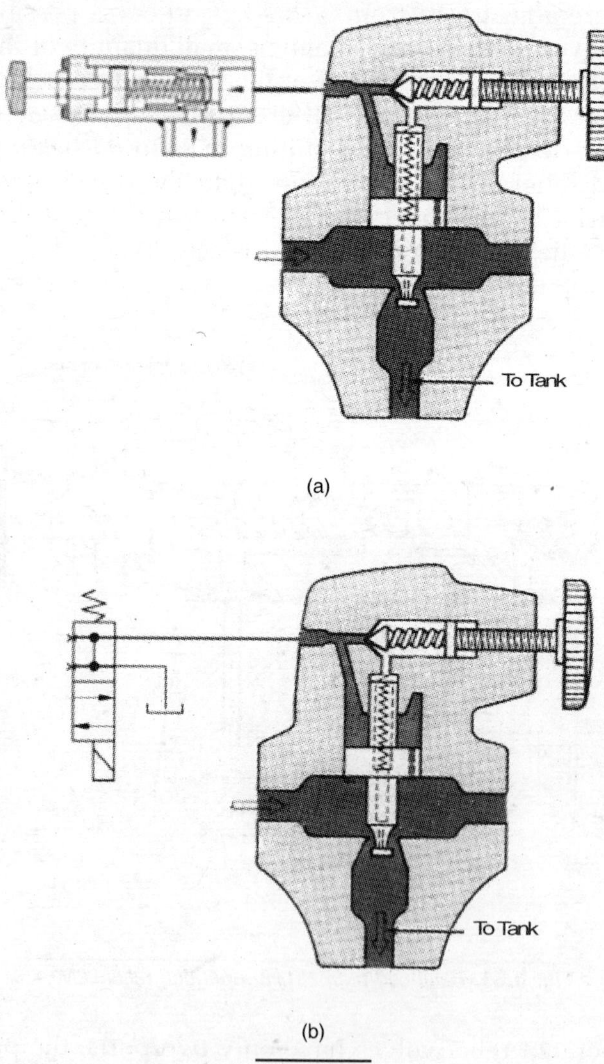

Fig. 5.32.

Pilot operated relief valves and solenoid valves are available as integral units. Such units are called *solenoid controlled pilot operated relief valves.*

Relief valves are available with threaded body ports that will facilitate their direct mounting on pipes. They are also available in sub-plate/ manifold block mounting configurations.

Pressure-reducing Valves. Pressure reducing valves are used maintain a reduced pressure at the outlet port as compared to the pressure at the inlet port. The pressure at the inlet port is controlled by the system relief valve and is set for a pressure higher than the pressure-reducing valve. If for instance, the pressure-reducing valve is set for a pressure equal to 100 kgf/sq-cm it will permit free flow from inlet to outlet so long as the pressure is below the valve setting. The spool itself remains in hydraulic balance because it is subjected to the same pressure on either side through an orifice in the center and the light spring is only holding the spool in a position that keeps the inlet and outlet ports wide open. When the pressure exceeds the valve setting the spool shifts to the left against the spring throttling the flow to the output line thus creating a pressure drop in that line. The drain line to tank ensures the valve is never fully closed by permitting a certain amount of fluid to bleed to tank.

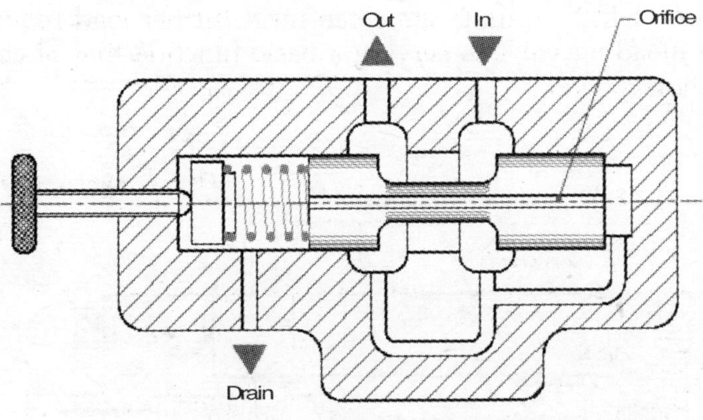

Fig. 5.33. *Pressure reducing valve (below valve setting).*

Direct operated pressure reducing valves of the type shown above cannot handle- large flows and are not very accurate either, in maintaining the desired pressures. Pilot operated pressure-reducing valves are recommended for large flows and accurate control. These are slightly modified versions of the pilot operated pressure relief valves. The main constructional features in both are otherwise similar. Pressure reducing valves are 'normally-open' spool valves. Free return flow is possible only where the return line pressure before the valve is less than the valve setting. If this pressure is higher a check valve for free return flow will be required. Pilot operated pressure-reducing valves with integral check valves are available.

5.5 UNLOADING AND PRESSURE SEQUENCE VALVES

Although they are used for different purposes a discussion on unloading and pressure sequence valves have been taken together because, their internal construction is almost similar. Perhaps, the only difference is that the pilot pressure that actuates the spool comes from an external source in case of unloading valves but originates from within the valve itself in case of sequence valves. An unloading valve is invariably used in two pump circuits to unload the bigger pump to tank at low pressure after the primary function has been achieved. Such circuits are commonly called 'hi-low' circuits where two pumps either as integral units or as separate units together feed the actuator to obtain a certain initial speed. When the actuator meets with load resistance, speed may no longer be the criteria and therefore the delivery from the larger of the two pumps is diverted to tank at low pressure using the unloading valve. The high pressure-low delivery pump alone can meet further load requirements. Here the unloading valve is serving a basic function that of conserving the input horsepower.

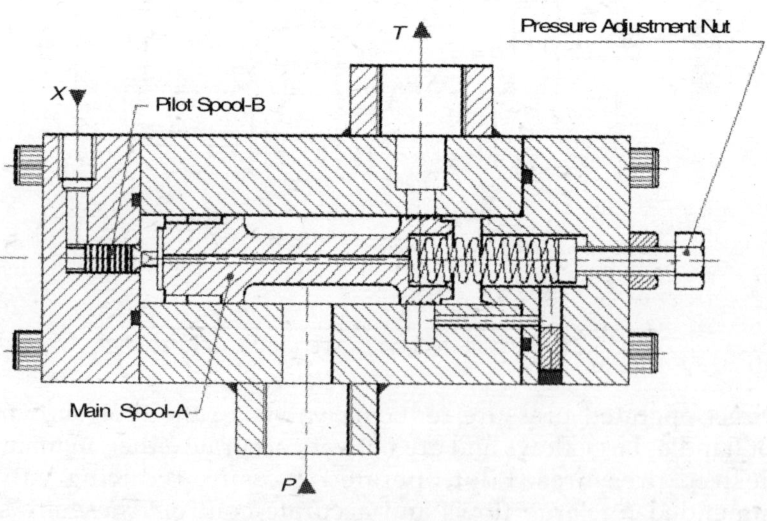

Fig. 5.34. *Unloading valve.*

Unloading valves are 'normally closed' type spool valves. This means the main spool *A* is keeping the tank port *T* closed until it is forced to uncover by application of external pressure on the pilot piston through port *X*. The low-pressure high delivery pump output line is tapped through a Tee connection and one line is connected to the *P* port of the unloading valve and the other combines with high pressure pump delivery as shown in Fig. 5.34.

(a)

Fig. 5.35. *Continued unloading and relief valve*

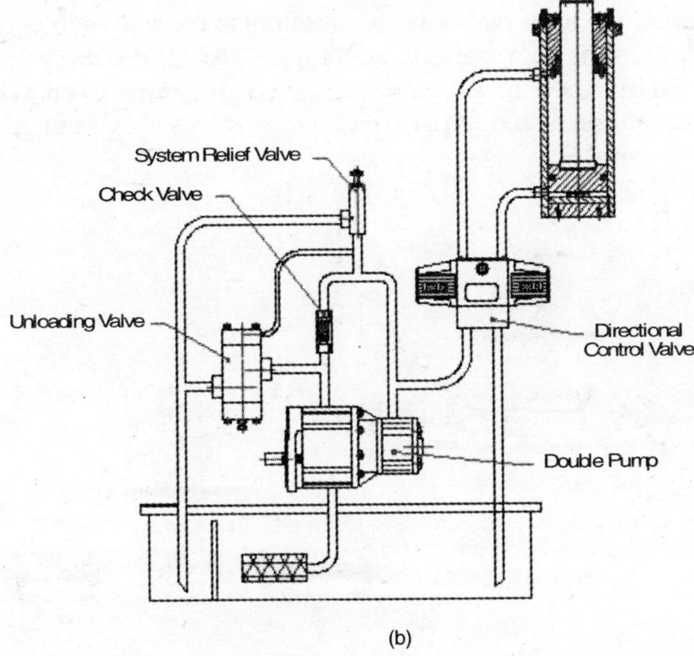

(b)

Fig. 5.35. *Unloading valve circuit.*

The main spool is biased against a spring that can be set for the pressure at which the valve should unload by using the pressure adjustment screw. These valves are internally drained and so no separate drain line to tank is required. Since the unloading valve unloads at low pressure the hi-flow pump out put, the main system relief valve may be sized to suit just the requirements of the high pressure-low delivery pump.

Integrated units that combine unloading, main system pressure relief and check valve functions in one block are also available. These valves not only unload the high-volume low-pressure pump but would also relieve the high-pressure low volume pump when the system pressure reaches the set value. No separate system pressure relief valve is required while using these valves. These units help conserve space and eliminate the need for individual piping. These dual function valves are also used in accumulator charging circuits to limit maximum system pressure and to unload the pump after charging the accumulator.

5.6 SEQUENCE VALVES

In many applications it is required that a certain operation must precede or follow certain other operation. Like for instance, in the example shown below the bending operation must precede the edge folding operation. Here, obviously we have two or more actuators to control, each performing different operations in a definite sequence. Also, the primary actuator has to be maintained at a pre-determined pressure even while the secondary operation is taking place. Sequence valves are designed to meet such requirements.

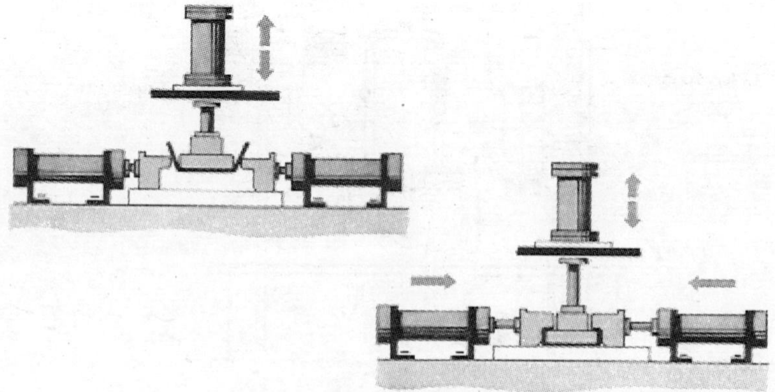

Fig. 5.36. *A sequencing operation.*

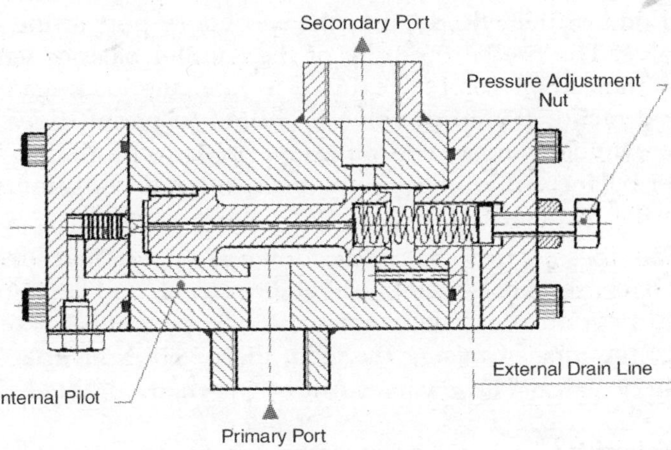

Fig. 5.37. *Sequence valve.*

A sequence valve differs from an un-loading valve in that the pilot pressure signal originates from the primary port of the valve itself instead of, from an external source. The un-loading valve if modified to this extent, can be made to perform as a sequence valve. The tank port in this case becomes the secondary port. As the pressure signal to the pilot spool is obtained from the primary port, the pressure at the primary port is maintained. Since both secondary and primary ports are under pressure during sequencing operation a separate external drain line to tank is required as otherwise the main spool would not respond to pilot pressure due to fluid collection on the spring side. Sequence valves are also available which respond to an external pilot pressure source.

Some designs are internally drained others need an external drain line to tank. They are available with a wide range of pressure ratings.

If the sequence valve is installed between the directional control valve and the primary actuator a separate bypass line to tank for the return flow is required in the form of a check valve. Sequence valves with integral check valves are available. A check valve is not necessary if the sequence valve is installed between the pump and the directional control valve.

5.7 COUNTER BALANCE VALVES

Counter balance valves offer free flow in one direction and block reverse flow up to a pre-determined pressure, which is infinitely adjustable within the specified range of the valve. Counter balance valves are used to achieve a controlled descent of an over-running load. They prevent the load from falling freely. The primary port of the counter balance valve is connected to the rod end of the cylinder and the secondary port to the directional control valve. The pressure setting of the counter balance valve should be slightly more than what is required to hold the load against gravity. When the directional control valve is shifted to permit flow to the cap end of the cylinder the rod end oil exerts pressure on the pilot piston of the counter balance valve causing the main spool to shift thus permitting flow from the rod end to tank.

The load does not drop but comes down at a controlled speed because a constant backpressure is maintained at the rod end and tank line pressure exists only beyond the directional valve. To permit reverse free flow required at the time of raising the load, these valves invariably have an integral check valve. These valves can be internally drained.

5.8 SHUTTLE VALVES, EXHAUST VALVES AND FLOW DIVIDER VALVES

Shuttle valves are three port ball check valves. The ball shuttles between any two ports *A* or *B* keeping one or the other closed depending upon whichever, of the two ports is at a higher pressure. In the valve shown below flow is permitted between $A \rightarrow C$ or $B \rightarrow C$ and $A \rightarrow B$ are blocked. Horizontal mounting is recommended for these valves.

Exhaust valves are hand lever operated puppet type valves, which permit free flow from port *A* to port *B* and blocks return flow in the normal position. By operating the lever the position can be shifted to permit rapid exhaust of compressed oil. They are also available with de-compression features for smooth exhaust. They are usually installed directly on a branch line tapped from the port that needs quick exhaust bypassing the directional control valves.

Fig. 5.38. *Shuttle valve.*

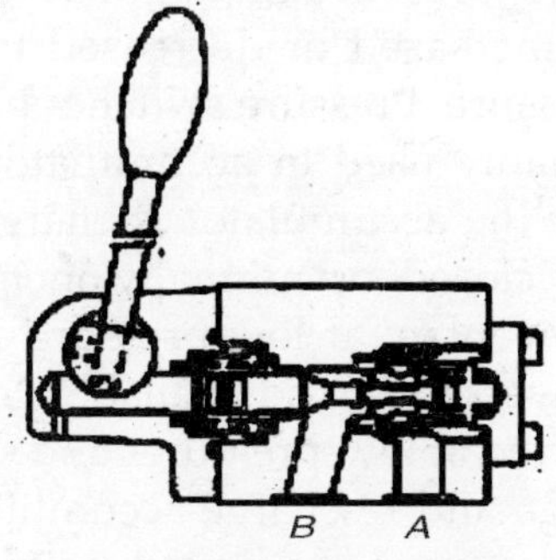

Fig. 5.39. *Exhaust valve.*

Flow divider valves are normally used to synchronize the speed of two or more actuators of equal piston areas served by a common pump. Synchronization is required because of varied resistance to motion individually offered by them due to inconsistent manufacturing accuracies.

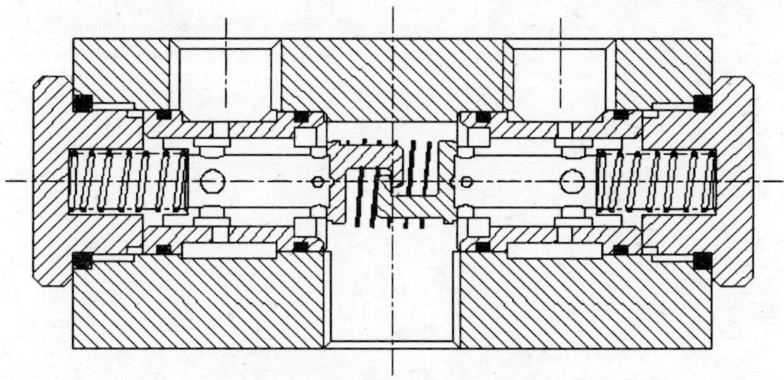

Fig. 5.40. *Flow divider valve.*

Inlet flow passes through two matched orifices one in each spool is connected to the two outlet ports. If flow to one outlet tends to increase, the greater pressure drop across the spool tends to shift the spool. This constricts the passage area, restricting the flow and thus equalizing the speeds.

5.9 PRESSURE SWITCHES

A snap action electrical micro-switch and a plunger which moves against a spring force together constitute a simple pressure switch. When the hydraulic system pressure is applied on this plunger it overcomes the spring force and makes contact with the micro switch thus triggering an electrical action. It has a pressure adjustment facility which means the spring tension can be increased or decreased to enables the contact to occur at the desired pressure. Pressure switches basically control, solenoid valves. They are invariably used in accumulator circuits to de-energies the solenoid valve after the accumulator is charged. The switches can be assembled for 'normally closed' or 'normally open' open contacts. Pressure switches that make the contact at low-pressure and break the contact at high pressure and vice-verse are also available. Such switches will enable electric motors to be started at low-pressures and stopped at high pressures and automatically charge and re-charge accumulators. They are available as sub-plate mounted units, pipe mounted units or face-mounted units.

6

POWER OUTPUT DEVICES: HYDRAULIC ACTUATORS

6.1 DEFINITION

They may be either *linear actuators* or *rotary actuators*. The pump output in the form of energy contained in the confined fluid is finally converted here into either a linear motion of a piston and rod assembly or rotary motion of a shaft depending upon whether the output device is a Hydraulic cylinder or a Hydraulic motor. The hydraulic circuit design begins with the choice and the sizing of an appropriate actuator.

6.2 HYDRAULIC CYLINDERS (LINEAR ACTUATORS)

Hydraulic cylinders are either *double acting* or *single acting*. In double acting cylinders both the forward motion (advance) and return motion (retract) is caused by pump flow admitted to either end of the piston, alternately. In single acting cylinders only the forward motion is caused by oil pressure, the return motion is due to either the spring force or due to force of gravity. Gravity returned cylinders are always cylinders with large diameter pistons also performing the function of piston rods.

These are usually called *Rams*. Single acting cylinders with small diameter pistons are almost always spring returned because the piston weight will not be sufficient to over come seal friction and return to its original position merely assisted by gravity.

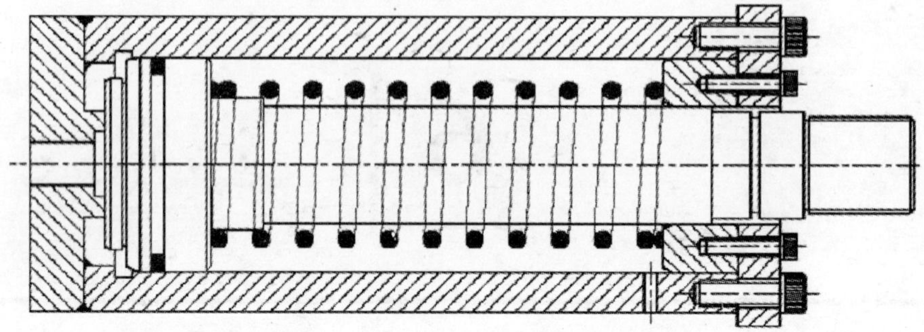

Fig. 6.1. *Single acting spring returned cyclinder.*

Standard *double acting cylinders* are also called differential cylinders because of the difference in areas that exists between the *rod end* and the piston end, also called the *cap end*. The rod end area is smaller to the extent of the area of the piston rod. Due to this variation the force that the piston can develop is more, when the piston advances than when it retracts.

However for the same reason the piston velocity during retraction is higher. If it is required that the piston must travel with the same velocity both during forward and reverse motions then we must use a cylinder with piston rods at both ends. Such a cylinder has the disadvantage that the force it can develop gets reduced to the extent of the rod area. Also, the overall length of a *double end rod* cylinder is more to the extent of 70 percent.

Consider a hydraulic cylinder which has a rod area equal to one half of piston area, that is, when $d = 0.71\ D$. If we admit oil to both the cap end and the rod end of such a cylinder simultaneously, although the pressures would equalize on both sides the piston would still advance because the forces are not equal due to difference in areas. But the retraction speed in this case will be equal to the approach speed because the rod area is made equal to 50 percent of piston area. This is one of the methods employed to equalize speeds of a linear actuator in both the directions without increasing the overall dimensions. The thrust or the pushing force available is however compromised to the extent of 50 percent. In this type of application the piston rod usually, is held stationary and cylinder is anchored to the load and is free to move.

Unlike other hydraulic components like the pump, the seals, valves and even the rotary actuators, hydraulic cylinders are not always

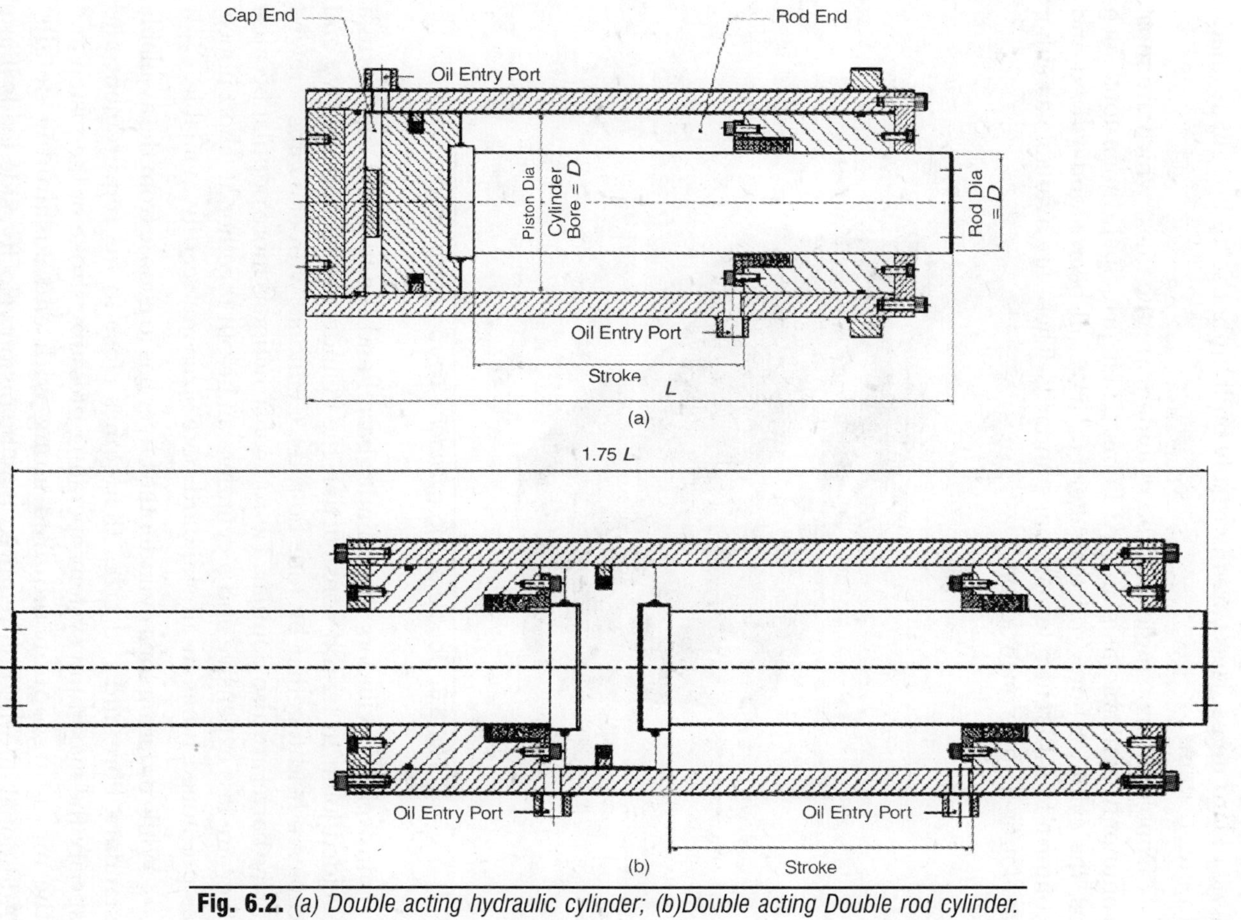

Fig. 6.2. *(a) Double acting hydraulic cylinder; (b)Double acting Double rod cylinder.*

'bought-out'. Many a times they are built 'in-house'. This is because they are simple to conceive and design and are amenable to batch production. Large single acting hydraulic cylinder and ram assemblies are not available as standard items. These are invariably built 'in-house' or bought as 'made to order' items.

Double acting, single end rod cylinders are the ones, which are most commonly available as standard items. Their principal dimensions such as, the cylinder bore, rod diameter etc, have all been standardized and conform to different international standards. These are available as either *welded* or *tie-rod* construction.

Fig. 6.3. *Double acting cylinder – tie rod construction.*

In welded cylinders the cylinder barrel and the two end caps form a welded joint. In tie-rod construction the cylinder barrel and the two end caps are held together by four tie-rods. Cylinders with end caps screwed to the barrel are also popular. Tie-rod construction is more popular because they are less expensive and a cylinder can be put together at 'short notice' as pre-honed tubes in all standard bore sizes are readily available.

While ordering for a cylinder the buyer has to choose from the available standard bore and rod sizes that comes close to his requirement and specify the maximum system operating pressure, stroke of the piston and the mounting option required along with end cushioning details. Additional information required may be piston velocity, side loads if any and the operating temperature. Bore sizes generally conform to ISO R1939 standards, all mountings and overall dimensions to CETOP –RP 58 H standards and port and piston rod sizes to ISO R2091.

A hydraulic cylinder needs to be anchored both to the frame and to the load. Here the designer has a variety of options to choose from. By choosing an appropriate mounting option he can ensure rigidity and sometimes altogether eliminate eccentric loads on the cylinder, piston rod and anchoring bolts. Eccentric loads on the cylinder or on the piston lead to early failure of seals and binding of piston. Generally a cylinder is anchored to the frame and held stationary whereas the piston rod is anchored to the load either directly or through end couplers such as a clevis or a yoke.

A hydraulic cylinder may be required to pull or push a load or for both of these applications. The piston rod and consequently the cylinder may be subjected to tensile, compressive, bending and buckling forces either severally or jointly. Also it may be required that the rod is held stationary instead of the cylinder. The choice of the mounting option depends on an analysis of these factors. Fig. 6.4(a) depicts a cylinder with a four bolt *front flange* mounting facility. This type of mounting is recommended for applications where the piston rod is not subjected to buckling load. This happens when the maximum force is developed on the rod end of the cylinder instead of at the cap end. This is generally the case when the cylinder is pulling the load. The reaction forces during such applications relieve loads on the flange and on the flange bolts.

When a cylinder is required to push a load, the piston road is subjected to compressive or buckling forces and *rear flange* mounting Fig. 6.4(c) is recommended for such applications as the reaction forces nullify the load on the mounting flange bolts and on the flange itself. Both rear and front flange mounting provide excellent centerline support for the cylinder. Because these are rigid mounting arrangements, it has to be ensured that the alignment between the cylinder and the load to which the rod is anchored is within permissible limits.

Rear end clevis, rear end fixed eye and Trunnion mounts [Fig. 6.4(b), (d), (i) and (j)] are the mounting options if the load doesn't move on a straight line but follows a curved path. They take care of misalignment in one plane.

Side lug and foot mounting [Fig. 6.4(f) and (g)] are almost similar and one or the other mounting methods can be used depending on space constraints. Both the methods impose a turning moment on bolts in addition to shear loads and this causes uneven bolt loading. This could be prevented to some extent by providing a butting face prevent the cylinder from sliding.

Centerline lug mounting (Fig. 6.4(h)) overcomes all of these problems because the lugs placed on the centerline of the piston rod. In this case all the four bolts are equally loaded and are subjected to only shear stresses.

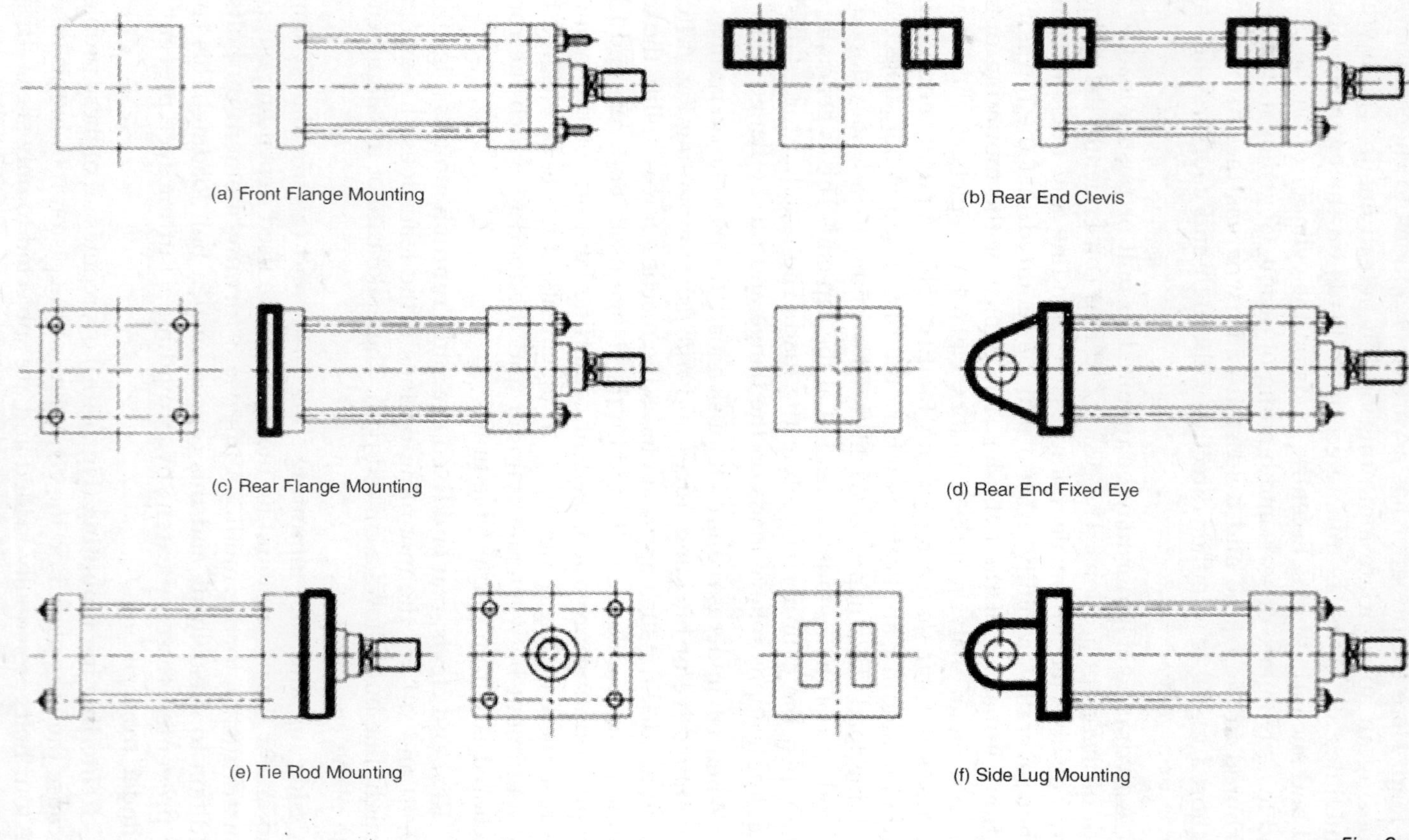

(a) Front Flange Mounting

(b) Rear End Clevis

(c) Rear Flange Mounting

(d) Rear End Fixed Eye

(e) Tie Rod Mounting

(f) Side Lug Mounting

Fig. Contd.

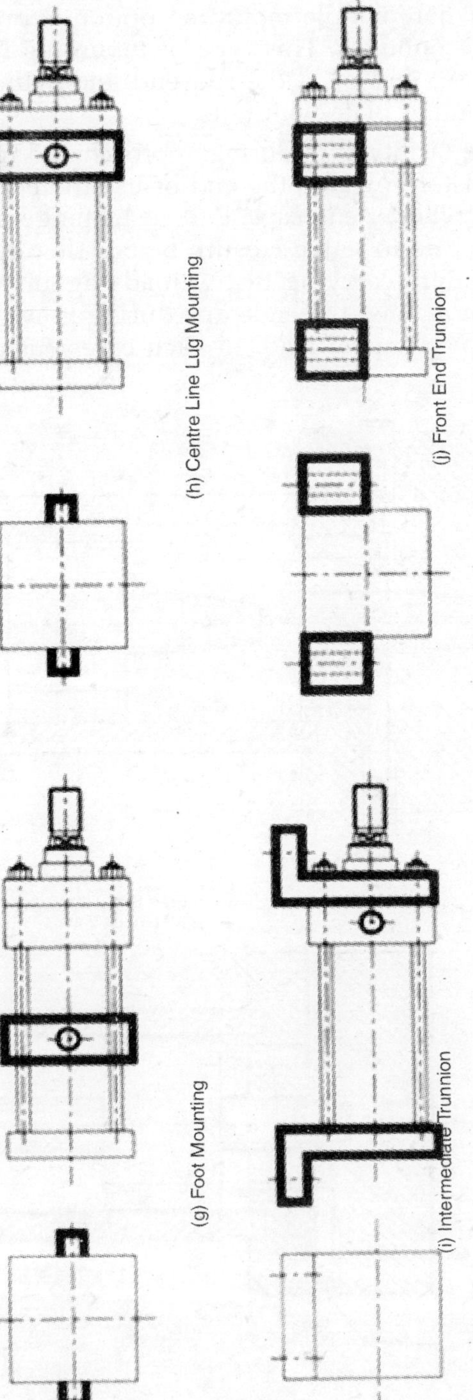

(g) Foot Mounting

(h) Centre Line Lug Mounting

(i) Intermediate Trunnion

(j) Front End Trunnion

Fig. 6.4

Tie rod mounting [Fig. 6.4(g)] is suggested only where small forces are involved and where simple mounting option is preferred to space constraints or for economics. This type of mounting has extended tie rods either at the cap end or at the rod end and cylinders are simply secured to the mounting plate with nuts.

End cushioning facility both during approach and return is required to decelerate the piston towards the end of its stroke and thus prevent impact load on the cylinder end caps. End cushioning may be eliminated if the pistons stop due to valve closure before they hit the end caps. Single acting cylinders carrying heavy loads against gravity have a tendency to bottom against the end caps during gravity fall and cause damage due to heavy impact load. In such cases cushioning becomes indispensable.

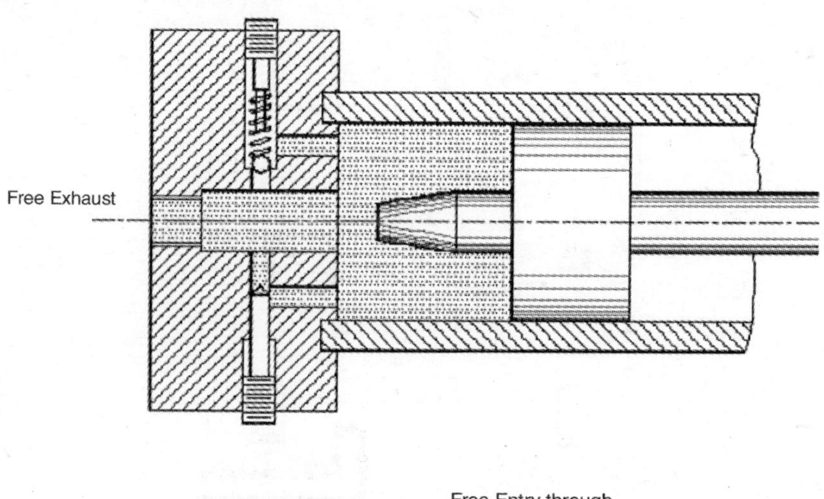

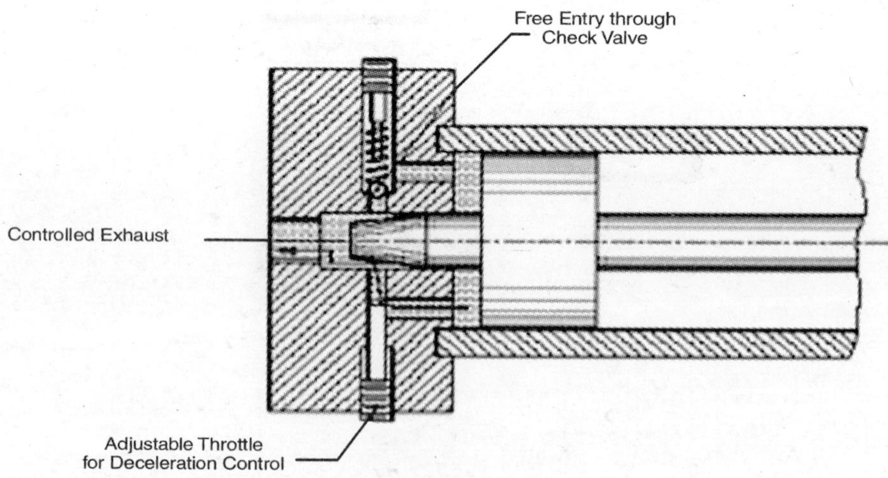

Fig. 6.5

Deceleration begins when the tapered plunger enters cap end and closes the discharge port forcing the oil to take the restricted path. The rate of deceleration can be controlled, by further restricting the flow passage through a needle valve. During rod extension the oil enters the chamber through a check valve.

The bore diameter of a hydraulic cylinder is determined based on the desired approach speed of the piston using the formula $Q = F/p3V_a$ where, Q is the required flow in cm^3/sec, V_a is the desired approach speed in cm/second, F is the hydraulic force that needs to be developed in kgf, and p is the maximum system operating pressure in kgf/sq-cm.

After obtaining the value of Q from the above formula cylinder diameter 'D' can be obtained from the equation,

$$D = \leq (Q/0.7854\ V_a)\ \text{cm} \tag{1}$$

The value of Q obtained represents the theoretical pump discharge in cm^3/sec. Since even the best pump has a volumetric efficiency no more than 95% the actual discharge Q_A must be taken = $Q/0.95$. Therefore Eq. (1) becomes

$$D = \leq (Q_A/0.7854\ V_a)\ \text{cm} \tag{2}$$

The diameter D so obtained must be rounded of to the nearest standard size. The manufacturer of a standard cylinder would have sized the piston rod diameter to resist buckling load based on the maximum thrust that the rod is subjected to. The buyer has no option but to select the size offered.

If the cylinder is manufactured 'in-house' then the diameter of the rod is determined by evaluating the maximum buckling load based on the 'Euler' equation treating the piston rod as a column subjected to buckling.

$$\text{Buckling load } K = (\pi^2 . E . I)/ L^2 \tag{3}$$

where E is the modules of elasticity for steel = 2.3×10^6 kgf/sq-cm, I is the second moment of area = $0.05\ d^4$ cm, (d = piston rod dia) L is the free buckling length in cm, equal to L itself if both the ends are pivoted and rigidly guided. $L = 2L$ if one end is free and one end is rigidly fixed. $L = 0.7\ L$ if one end is rigidly fixed and the other guided and pivoted. $L = 0.5\ L$ if both ends are rigidly fixed and guided.

Piston rods are made out of case hardening steels which are case hardened to a depth of 0.5 mm to obtain a case hardness value 40–50 HRC and then finally ground to retain a case depth = 0.2. If piston rods are exposed to harsh environmental conditions they could be manufactured out of medium carbon steel and then hard-chromium plated to a thickness of 40μm (0.04 mm). To obtain proper metal deposition and smooth finish the piston rod must be ground both before and after plating. The final finish obtained should be between R_a= 0.1 to 0.15 μm.

The buckling load K represents the maximum piston thrust at which the rod tends to buckle. To obtain a rod diameter that resists buckling, a factor of safety ranging from 2.5 to 3.5 must be applied to this load K, to obtain the safe load K_S. Therefore $K_S = K/\text{factor of safety}$.

After obtaining the value of piston rod diameter d from equation (3) we now proceed to determine the piston velocity in cm/sec during return stroke based on the equation, $V_R = Q_A/0.7854\ (D^2 - d^2)$ where D is the piston diameter also considered equal to cylinder bore diameter and d is piston rod diameter in cm.

The stroke volume of the cylinder in liters during approach = $(0.7854\ D^2 s)/1000$ and during return = $\{0.7854\ (D^2 - d^2)\ S\}/1000$ where S is stroke in cm.

6.3 DESIGN OF CYLINDER BARREL

The cylinder is subjected to both the hoop stress and the longitudinal stress. Since longitudinal stress is one half of hoop stress and is often balanced by the applied load, it is not considered while determining the cylinder wall thickness. As the cylinder must not have a permanent set, the allowable limit is the limit of proportionality. This limit for mild steel is equal to the yield stress and about 90% of yield stress for alloy steels. The safe working stress is limited to 20 – 35% of yield stress, the lower value being considered in the presence of stress raisers such as a hole, groove, thread or weld. The hoop stress σ as given by the equation $PD/2t$ gives fairly accurate results up to a pressure of 300 kgf/sq-cm. Beyond this limit the Lame's equation for thick cylinders must be used.

The cylinder wall thickness $t = PD/2\sigma$, where t is the desired wall thickness in cm, P is the maximum operating system pressure in kgf/sq-cm, D is the cylinder bore diameter in cm., and s is the safe working stress in kgf/sq-cm. Safe working stress σ lies between 1200 to 1500 kgf/sq-cm, for available medium carbon seamless steel tubes. The cylinder bore is ground and honed to obtain a surface finish R_a = 0.16 – 0.08 μm (0.016 – 0.008 mm).

If the cylinder is of 'tie-rod' construction the force on the 'tie-rod' is the force F developed by the cylinder at maximum system working pressure divided by number of 'tie-rods'.

The diameter of the tie-rod is given by

$$d = \sqrt{F / n \times \sigma_t \times .7854} \tag{5}$$

where, n is number of tie-rods, σ_t is the maximum permissible tensile stress for the tie-rod material in kgf/sq-cm, F is the maximum force developed by the cylinder in kg. Tie-rods are usually made out of high tensile alloy steels. It is important that the tie-rods are tightened to a torque level that pre-loads them to a force, which is in excess of the maximum force exerted by the cylinder.

The relation between the applied torque T in kgf-m and the resulting tie-rod pre load F_p in kgf is given by $(F_p . d)/T$ = constant where d is the core diameter of the thread in meters, F_p can be taken equal to the maximum force exerted by the cylinder and the constant can be taken equal to 5 on the average but lies between 3–10 depending upon the smoothness of the surface the nut is in contact. Higher the smoothness, higher will be the value of this constant.

Piston rods subjected to heavy side loads require proper bearing support. The length to diameter ratio of the bush in which the rod slides is an important design criterion. This ratio should be greater than 2. The material for bush is either phosphor bronze or bronze/PTFE compound.

Piston can be screwed to the piston rod with *O* ring placed in the joint interface or the piston and the piston rod could be welded in semi-finished condition and then machined in one setting. Piston is manufactured out of medium carbon steel (E_n 8/*C*40) or it could be a steel casting or S.G. iron casting. Seal locating grooves must be machined on the piston strictly to the recommendations of the seal manufacturer.

In cases where the piston rods are subjected to heavy side loads, it is important to prevent the piston from extending up to the rod end cap in order to provide sufficient bearing length for the piston to resist side loads. *Stop tubes* of sufficient length inserted over the piston rod acts as a spacer and maintains the minimum distance between the piston and the bushing thus reducing the reaction load.

Large diameter cylinders and rams are usually made out of alloy steel castings or S.G. Iron castings. The safe working stress for these materials is limited to 25% of the ultimate tensile stress. Piston can be cast hollow on one side to reduce weight if sufficient cross sectional area is available to resist crushing forces. The seal groove in large diameter single acting cylinders is invariably located in the cylinder and the ram is hard chromium plated and ground.

6.4 CYLINDER EXPANSION DUE TO HOOP STRESS

When the fluid in the cylinder is pressurised the cylinder expands due to pipe wall elasticity producing an effective bulk modulus less than that of the oil under compression. This expansion has a direct effect on the seals contained in the piston. The seals tend to extrude in to the more than desired clearance that is formed due to this expansion. The extrusion gap varies with pressure and must be taken in to account in high-pressure applications where non-metallic seals are used.

Also, in applications where a static load must held for quite some time under zero leakage conditions it is very important to avoid the excess clearance that results due cylinder expansion. This diametric expansion δ is given by $\delta = d\,\sigma/E$, where δ is diametral expansion in cm, d is cylinder bore in cm., σ is the hoop stress in kgf/cm^2 and E is the modulus of elasticity for steel = 2×10^6 kgf/cm^2.

For thin wall cylinders having a wall thickness-to-bore ratio ≤ 0.1 hoop stress is given by the equation $\sigma = pd/2t$ and for cylinders having a wall thickness to bore ratio greater than 0.1, $\sigma = p\,\{(K^2 + 1) / (K^2 - 1)\}$.

Here, p is the pressure to which the cylinder is subjected, t is the cylinder wall thickness in appropriate units and $K = D/d$ where D is the outer diameter of the cylinder and d is the bore diameter.

6.5 UN-CONVENTIONAL CYLINDERS

There are a few variations to the conventional cylinders discussed above. These can be grouped under 'un-conventional' cylinders. These are, Rod-less cylinders, magnetic cylinders and telescoping cylinders and duplex cylinders.

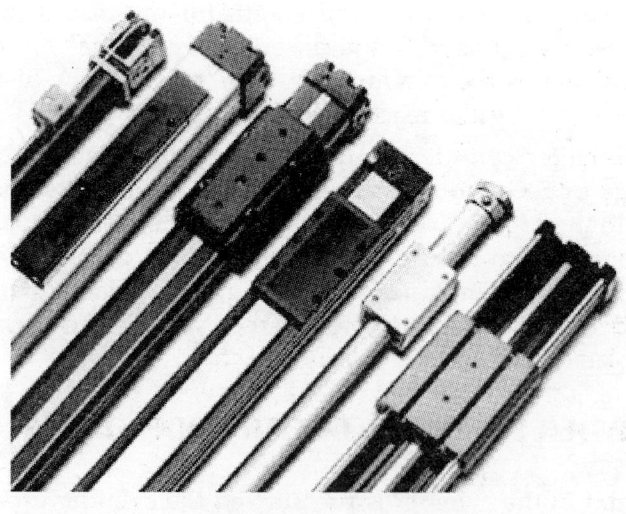

Fig. 6.6. *Rodless cylinder.*

Rod-less cylinders are identified by the absence of a piston rod at either end. They have only a piston contained within the cylinder. The piston is free to move from one end to the other and is connected to the external load by a lug protruding through a slot in the cylinder. Rod-less cylinders for the same stroke save installation space to the extent of 50 percent, as compared to conventional cylinders.

Magnetic cylinders are similar to rod-less cylinders but they differ in the way the piston is coupled to the load. Instead of direct coupling through a lug the two are coupled magnetically.

Telescoping cylinders are used where exceptionally long strokes are desired with comparatively small collapsed length due to limited space available for mounting. Fully extended length can be as much as four times the collapsed length. Force output varies with rod extension. It is highest at the beginning of the stroke where full piston area is exposed and is lowest at the end of the stroke where only the small piston area is exposed.

Fig. 6.7. *Telescoping cylinders.*

Both single and double acting versions are possible. Single acting versions are the ones widely used and they extend due to fluid pressure and retract due to opposing mechanical force. These cylinders are used in large forklifts and dump trucks.

These cylinders are constructed in stages using drawn over mandrel steel tubes which fit successively into each other. Stage, refers to a moving member and up to six stages are possible. The main cylinder that does not move is not counted as a stage. As the pressurized fluid enters the main cylinder the stage with the largest diameter extends first because the force component is maximum. Once it reaches the end of its stroke the next stage extends.

For force, flow and speed calculation purposes the effective diameter of a single acting-telescoping cylinder at any stage is the outside diameter of that stage. While speed is the sum of the speeds of each stage the force to be developed should be calculated based on the area of the last stage plunger. The product of this area and the maximum working pressure indicates the ultimate lifting capability of the telescoping cylinder. The total volume of oil required to fully extend the cylinder is obtained by dividing the total stroke by the number of stages and by multiplying the quotient by the effective area of each stage. The sum of this volume gives the approximate volume of oil required to fully extend the cylinder. The flow rate 'Q' required to achieve the total extension within a certain time is the sum of the flow rates for each stage.

6.6 DUPLEX CYLINDERS

A duplex cylinder comprises of two or more double acting cylinders mounted in line with pistons unconnected. The cylinders may be mounted so that the rod extend in the same direction or may be mounted back-to-back. The cylinders are often used to provide multi-position operation. A three stage duplex cylinder can provide up to four piston rod positions.

6.6.1 Synchronising the Simultaneous Motion of Two Cylinders

Two cylinders if they have to move together must have rigid mechanical linkage. Otherwise it is difficult to get them to move in unison because of variations in cylinder bores due to variations in rate of fluid leakage and seal friction. The cylinders will move together if both the cylinders can accept the same amount of fluid, which is difficult to ensure.

In cases where it is not possible to provide a rigid mechanical linkage between the two cylinders various other alternatives are adopted. *Rack and pinion* arrangement as shown in Fig. 6.8(a) can provide the solution if backlash can be tolerated. This method is however limited to cylinders which move in the same direction. **Series piping of cylinders** Fig. 6.8(b) is another method, which synchronises the movement of two cylinders fairly accurately if double acting double rod cylinders are used. This of-course is under the assumption that the system can accommodate the additional space requirement imposed by the double rod cylinder.

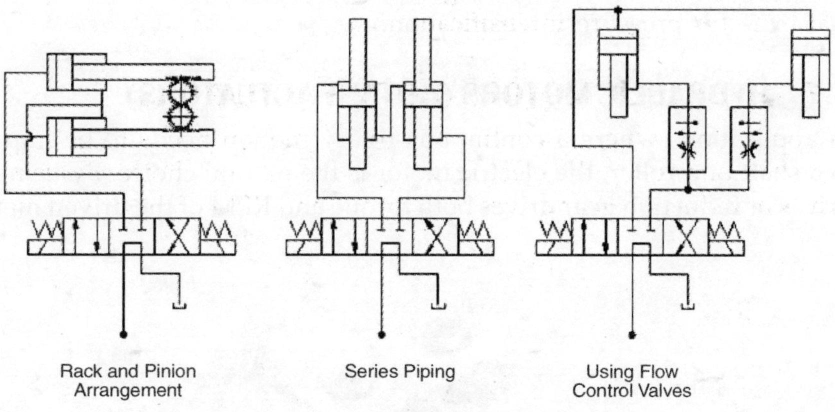

Fig. 6.8

Pressure and temperature compensated flow control valves Fig. 6.8(c) can ensure accurate synchronisation of two cylinders since the flow is infinitely adjustable and each valve acts independently of the other.

Two or more hydraulic motors coupled together in parallel Fig. 6.8(d) can act as rotary flow dividers capable of dividing one input flow into as many output flows as there are motors and can meet the flow demands of each cylinder thus regulating their speeds and hence their synchronisation. If both the cylinders have equal areas then both the motors must have the same displacement.

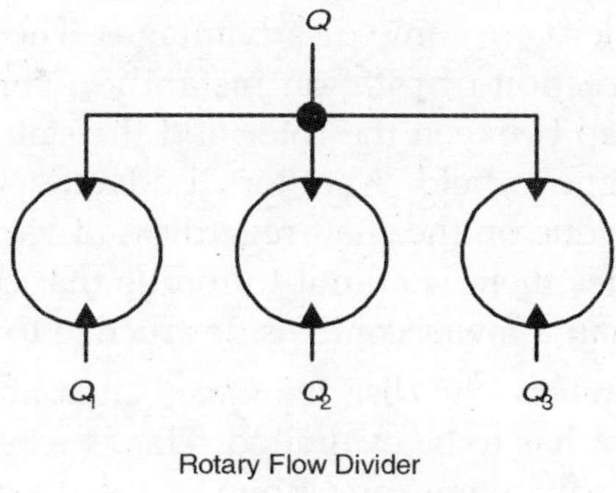

Fig. 6.9. *Rotary flow divider.*

Since all the motor sections turn at the same speed out put flows are proportional to the sum of the displacements of all the motor sections.

Rotary actuators can handle larger flows than flow divider valves also used to synchronizing the movement of two cylinders. When specifying rotary flow dividers design engineers must bear in mind the potential that exits for pressure intensification.

6.7 HYDRAULIC MOTORS (ROTARY ACTUATORS)

In applications where a continuous rotary motion needs to be imparted to a shaft or a roller, the electric motor is the natural choice. By using belt drives or reduction gear drives both torque and RPM of the driven member

Fig. 6.10. *Hydraulic motor.*

can be increased or decreased. By using variable speed drives it possible to infinitely vary the speed. Electric motors as prime movers are indispensable but there are some disadvantages. They cannot be instantly stopped or their direction of rotation instantly reversed. This is because there exists an air gap between the rotor and the stator and what connects them is a weak magnetic field. Another disadvantage is that they cannot maintain constant torque on the shaft regardless of variation in the speed of rotation/RPM. Besides there is a limit to torque that can be developed and 'stalling' for more than a few seconds is destructive to the life of the motor.

In applications where the disadvantages cited above becomes critical an alternative system has to be evaluated. This is where hydraulic motors, which have all the advantages and none of the disadvantages of electric motors could become an ideal choice. Hydraulic motors can be stalled for any length of time, their direction of rotation can be instantly reversed and their rotational speed can be infinitely varied without affecting their torque and they can be braked instantly and have immense torque capabilities.

Hydraulic motors are a very compact unit compared to electric motors. For the same power they occupy about 1/20 of the space required for electric motors and weigh about 1/10 of an electric motor. The moment of inertia to torque ratio for an electric motor is 100 as against a value equal to 1 for a hydraulic motor. The external appearance and the internal engineering of a hydraulic motor are similar to the hydraulic pump about which we have already discussed. They are available as gear, vane or piston units.

A pump can be physically distinguished from a hydraulic motor only by noticing the subtle difference between the two. The hydraulic motor has both the ports of the same size where as in case of a pump the inlet port is bigger than the outlet port Functionally too the hydraulic motor is similar to a hydraulic pump except that the roles are reversed. A pump accepts low energy hydraulic oil at the in put end and delivers high energy hydraulic oil at the put end in response to high speed rotary motion imparted to its shaft by a prime mover such as an electric motor. Here, basically *work is done on the fluid*. A hydraulic motor on the other hand receives high-energy hydraulic fluid at the in put end that performs work on the rotor and causes shaft rotation regardless of load resistance. *Here work is done on the shaft.*

The fluid after performing work discharges through the out let port. The direction of rotation of all hydraulic motors is reversible simply by reversing the direction of oil flow. In linear actuators we are basically interested in the force that the cylinder can develop and the speed/velocity of the piston. In rotary actuators we are interested in the torque or the turning moment that the shaft is capable of exerting on the load and the minimum shaft speed or RPM. In either case it is the pump discharge that determines the desired shaft RPM of a hydraulic motor or the piston velocity of a hydraulic cylinder. Where linear motion is desired, we size the cylinder first, based on the approach speed required and select a pump to suit. Likewise, *where rotary motion is desired we have to size the hydraulic motor first and select the pump to suit.*

The size of a hydraulic motor is determined by its displacement capability q which is the quantity of fluid that the motor can displace in cubic centimeters per revolution. This is given by the equation $q = (2.\pi.T)/p$ where T is the required torque in kgf-cm that the hydraulic motor should develop p is maximum pressure that the pump can develop in kgf/sq-cm. The torque rate $T_R = T/p$ in appropriate units. Pump discharge Q in LPM required to serve the motor is given by $Q = (N \, . \, q)/1000$ where N is the maximum desired shaft RPM.

In the above equations a suitable value for the volumetric efficiencies of both the pump and the motor must be introduced to obtain practical results. Also *the starting torque required* is governed by the inertia of the driven load.

Therefore a (10 to 30)% increase in the value of T obtained is required to start a given load. If accurate value is desired it can be obtained from the equation $T_A = (J.\delta n)/t$ where T_A is the additional torque in kg-m required to start the load, J is the torsional polar modulus in kgf-m,^{2} δn is the change in speed in time interval t.

There are two types of hydraulic motors. *High speed-low torque motors and low-speed high torque motors.* In high speed-low torque motors the shaft is driven directly from either the barrel or the cam plate whereas in a low-speed high torque motor the shaft is driven through a differential gear arrangement that reduces the speed and increases the torque. Depending upon the mechanism employed to produce shaft rotation hydraulic motors can be, either gear motors, vane motors or piston motors. Gear motors are either external gear motors or internal gear motors. A variation to internal gear motor is the roller-Gerotor motor and the crescent gear motor. Piston motors could be axial piston motors or radial piston motors.

External gear motors are the simplest form of motors consisting of two spur gears mounted on two shafts each supported at either end in bushings.

The two spur gears are in mesh and rotate together. The shaft of one of the gears is extended through the housing and acts as the driving member. High-pressure oil entering the chamber applies hydraulic pressure uniformly on all the gear teeth exposed to oil pressure in that segment. Since pressures are equal on either side of the teeth the forces cancel each other out leaving only the last teeth to react to the force causing rotational motion. In other segments too, the net torque is zero making the *torque available a function of one tooth.* In the discharge segment the teeth are exposed to tank line pressure and hence there is no opposing force to the turning moment.

The advantages and disadvantages of a gear pump already discussed under the relevant chapter, applies equally to gear motors. They are cost effective and dirt tolerant. They have lower volumetric efficiencies compared to other motors. The maximum speed attainable is limited 2500 RPM and maximum pressure to about 140 kgf/sq. cm and can develop a maximum torque ≅ 65 kgf-meter. The above values could be higher for high-performance gear motors but their cost advantage is lost.

Internal gear motors are often called *gerotors,* a short for generated rotors. A gerotor consists of a stationary ring in which a lobed star is made to rotate with an eccentricity e with respect to the stationary ring. A widely used variation to gerotors is the roller-gerotor motor with differential gear arrangement used primarily for low speed-high torque requirement.

In this arrangement rollers replace lobes of the outer gear resulting in lower friction and consequently lower starting torque, improved mechanical efficiency and extended service life. These motors can with stand pressures as high as 300 kgf/sq-cm and develop torque as high as 150 kgf-m at speeds ranging from 5–900 RPM.

Crescent gear motor is a gear within a gear arrangement. These motors are good for high-speed low torque applications.

Vane motors develop their torque similar to gear motors but the force transmitting member in this case are the vanes, held extended against the cam ring by the oil pressure In some designs spring force may also assist in keeping the vanes held against the cam ring. The spring here, in fact functions like a rocker arm. One rocker arm is in contact with two vanes alternately forcing one vane to move out of the slot and bear against the cam ring even while permitting the other to collapse in to the rotor slot. Maximum torque capability for these motors is about, 45 kgf-m at a pressure of 150 kgf/sq-cm. Maximum speed capability is up to 3000 RPM.

Low speed high torque vane motors are available which operates at speeds as low as 5–150 RPM and can develop torque as high as 600 kgf-m. These motors are ideal for conveyors, turntables, mixer drives, winches etc.

Piston motors are available in three different configurations. Axial or inline piston motors, Radial piston motors and bent axis piston motors. Piston motors have the highest efficiency compared to other types. They are also capable of high speeds and high torques. Among them the inline piston motor is more popular in industrial hydraulic systems because of its simplicity and cost advantage. They have a reputation for high volumetric efficiency and are excellent for both high and low speeds. These pumps are good for pressures up to 350 kgf/sq-cm, torque up to 200 kgf-m and up to 4500 RPM.

Radial piston motors are the only choice for displacements in excess of 350 cc/revolution. They are also the ideal motors for low speed high torque applications. They provide high starting torque but they have a speed limitation, about 2000–2500 RPM is the design limit.

Hydraulic motors find their application both in industrial hydraulic systems and in mobile hydraulic systems. However they are more widely used in mobile hydraulic systems such as in cranes, winches, ships propulsion and steering and in hydrostatic transmission drives used on front-end loaders and trawlers. Because of their high torque and low inertia characteristics they are extensively used in servo-controls.

Fig. 6.11. *Radial Piston Motor.*

6.8 MOTOR SELECTION

Motors are selected based on the duty cycle. For output speeds in excess of 800 revolutions per minute it is common to use a high-speed axial piston motor. Low speed hydraulic motors are used where the out put speed required is less than 800 revolutions per minute. A low speed motor weighs twice as much compared to a high-speed motor of the pressure and power rating, but the weight of the gearbox required must be borne in mind.

A high-speed motor however has a price advantage of about 5 percent over low speed motor even after taking into account the additional cost of the reduction gearbox. Due to low rotational speed and due to the elimination of the reduction gearbox low speed motors are more reliable than high-speed motors. Radial piston motors are more reliable and are easy to maintain compared to other motors.

Torque generators are also a kind of rotary actuators but the rotational movement that the shaft can execute is always less than one complete revolution. Within this limited angle of turn they can perform both clockwise and anti-clockwise motions. Torque generators of the rack and pinion type are capable of more than one revolution, the limiting factor being the length of the rack.

Fig. 6.12. *Torque generator.*

Rack and pinion type of torque generators are nothing more than hydraulic cylinders with pistons at either end connected to each other through a rack. A pinion whose shaft acts as the out put device is in mesh with rack. The pinion housing is firmly mounted on the tie rods and is placed central to the cylinder. The cylinder has ports at either end for oil entry. If the oil is admitted to the left port the piston moves to the right and vice versa. Depending upon the direction of piston movement the pinion executes clockwise or anticlockwise motion. Out put torque depends on the force. exerted by the piston and therefore on the cylinder area and the working pressure. This torque can be doubled in a double rack arrangement.

7

AUXILLARY DEVICES

7.1 HYDRAULIC OIL SEALS

This chapter excludes any discussion on rotary shaft seals or lip seals which are primarily used for retaining lubricants in rotating shafts. They are good for very low pressure sealing applications and are seldom used in high-pressure oil hydraulic systems.

It is appropriate that any discussion on seals must begin with a few words on leakage of fluid medium, both internal and external. Small quantities of fluid escaping through the working clearances is either allowed or tolerated if the leak is internal and as long as it does not affect the integrity of the system. Internal leakage does not result in loss of fluid as the fluid is eventually returned to the reservoir. If the performance of the system is affected then it is an indication that the leakage has advanced beyond tolerable or allowed limits. This may be due to wear and tear of the rotating or reciprocating components or the sealing elements. In such cases only replacement of the worn out parts would solve the problem.

External leak on the other hand, is neither allowed nor tolerated. This is because it is expensive since the leaking oil can rarely be recovered. Also, it makes the work area messy and hazardous. External leak could be due to high operating pressures, poor design practices, improper installation and oil contamination.

7.2 STATIC SEALING APPLICATIONS

Seals are required to prevent oil leakage and maintain pressure. Efficient performance of any hydraulic system depends almost entirely on the SEAL, a component that represents a fraction of the total cost of the system.

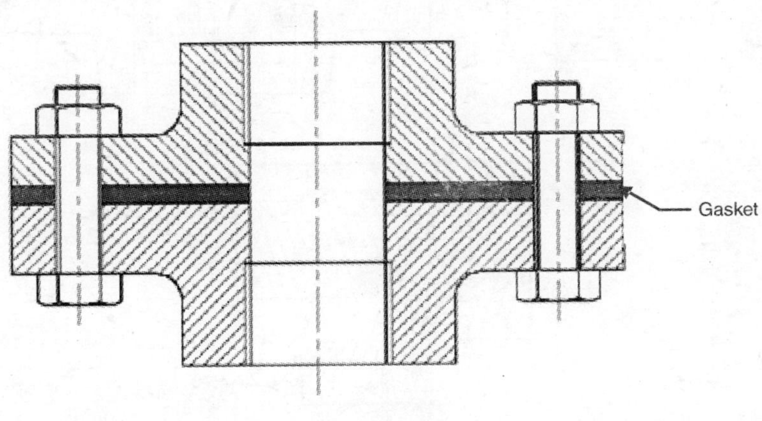

Fig. 7.1

A simple gasket inserted between two flanges tightened with bolts, effectively prevents leakage of fluid at pressures less than 5 kg/sq-cm and hence could be termed a seal. For effective sealing however the mating faces must have smooth finish and the gasket must be compressed uniformly with sufficient number of bolts. Here it should be noted that both the flanges and the seal are stationery and hence it could be called *a static sealing application.*

Another very effective but inexpensive seal is the *O*-ring, so called because it is round and has a cross section that is a circle. It is specified by its cross sectional diameter and inner or outer diameter depending upon whether it is located on the inner diameter or outer diameter. *O*-rings can be employed in both static and dynamic sealing applications. *O*-rings when used under static sealing conditions can withstand very high pressures and provide virtually a leak proof interface.

Dynamic *sealing application* on the other hand, involves a relative motion, which is either reciprocating or rotary, between the seal and its enveloping surface. *O* rings are not the recommended seals for this kind of applications because of the volume-swell of rubber due to fluid absorption, a property that is preferred for static applications and must be avoided in dynamic applications. Also, they will extrude into the working clearance even under moderate pressures and speeds and thus get damaged.

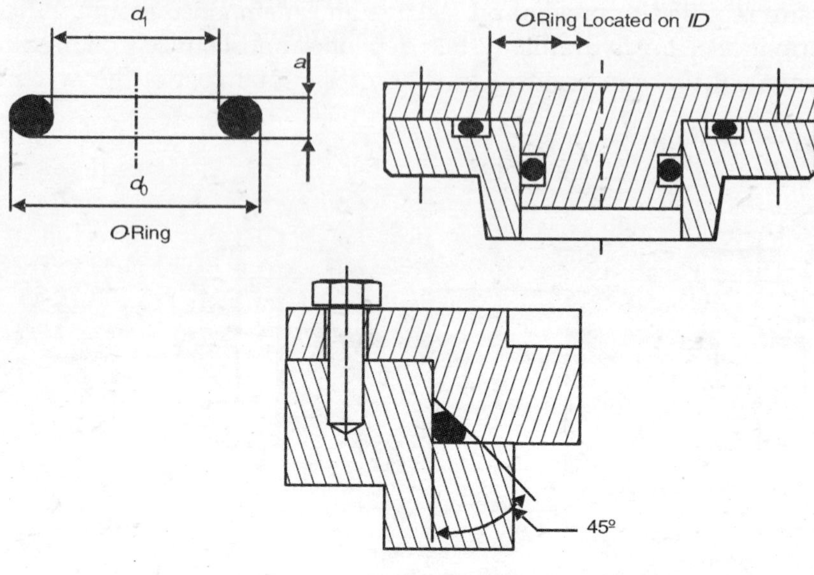

Fig. 7.2

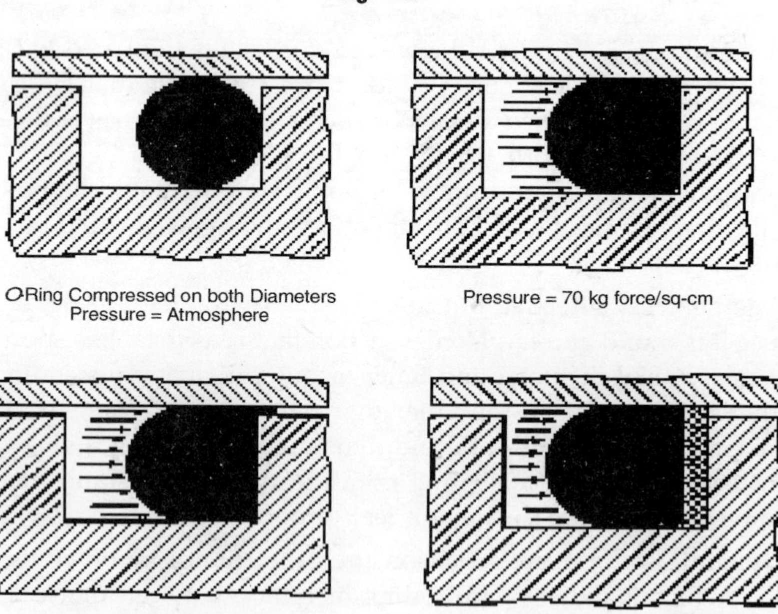

Fig. 7.3

The maximum recommended pressures for this kind of applications is about 50–60 kgf/sq-cm. Installation of PTFE back up rings would prevent such extrusions and enable the *O*-ring to hold higher pressures. The mechanism of extrusion of *O*-ring in to the annular clearance between the piston and the cylinder under dynamic applications is shown in Fig. 7.3. Quad rings are similar to *O*-rings but have a four lobed *X* type cross section with annular grooves instead of round cross section. For dynamic applications they are better than an *O*-ring because they require less squeeze to maintain the sealing effect and therefore have lower frictional resistance. The four lobed construction resists the tendency to roll in the groove when used as a reciprocating seal. Spiral twist and extrusion problems are less severe. They fit well in the standard 'O' ring grooves.

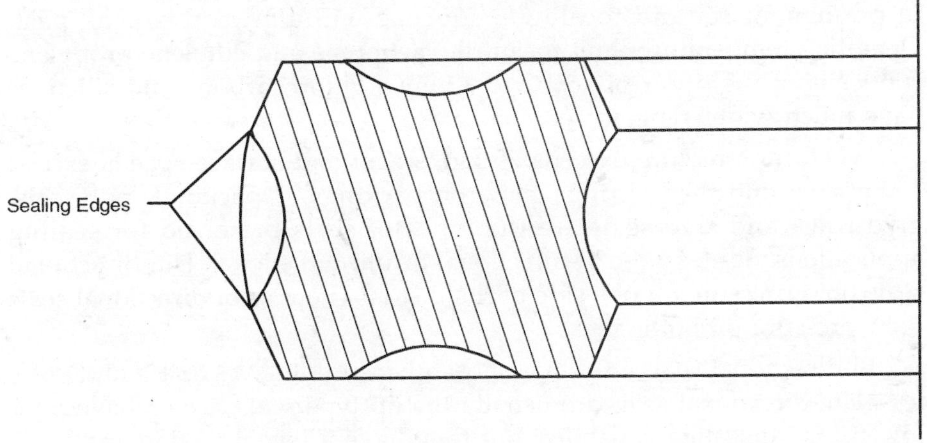

Fig. 7.4. *Cross section of quadring.*

O-rings are precision molded seals manufactured out of synthetic rubber since natural rubber is not oil resistant. The most popular materials used in oil hydraulic systems are, nitrile rubber (NBR), Chloroprene or neoprene rubber, ethylene propylene, fluorocarbons such as viton, and silicones. Among these nitrile and neoprene rubber are the most common. Hardness, tear resistance, permanent set and resistance to temperature are the qualities considered as important in any elastomer intended for sealing applications in oil hydraulic systems.

Hardness is an index of resistance of any elastomer to deformation and is measured by pressing a ball or a blunt object on to the surface of rubber. The most commonly used instrument is the durometer made by the shore instrument company. On shore *A* scale, 0 is considered as soft and 100 as hard. Hardness of most elastomer seal materials is in the range of 40–90 shore *A*. Within this range, the variation in hardness can be achieved through compounding. The specified hardness for *O*-rings is 70-shore *A*.

Tear resistance is a measure of the stress required to rupture a sheet of rubber after an initial cut. The seal can fail if it is accidentally scratched during installation if tear resistance is poor. Most elastomers have only moderate tear resistance and therefore demand proper installation techniques.

Permanent set is the deformation of an elastomer that remains after removal of the load causing the stress. Most elastomers have very good resistance to compression set. The maximum compression to which the *O*-ring may be subjected to, during installation is about 30 percent of the *O*-ring cross sectional diameter. Over Compression or squeeze will result in permanent set and too low a squeeze will allow oil to leak past. Operating temperature limit for nitrile, neoprene and ethylene-propylene rubber is of the order of –55ºC to +145ºC. Fluorocarbons and silicones have much wider ranges.

When the pressure exceeds 70 kgf/sq-cm and surface speeds exceed 100mm/second then *U*-rings, chevron packings/V packings, polyseals, glyd-rings and step seals are the types of seals preferred for sealing applications. Seals can be bi-directional or uni-directional. Uni-directional seals hold pressure on one side of the piston where as bi-directional seals hold pressure on both sides.

Bi-directional seals are often chosen because it saves space and costs less. Uni-directional seals are usually the lip type seals, which have poor low pressure sealing capability but they have a lower friction and wear rating compared to bi-directional seals, which are essentially, the squeeze type and therefore have higher friction and wear rating. Uni-directional seals are preferred in single acting cylinders and in rod sealing applications. Bi-directional seals are more difficult to install in their grooves and require special tooling. Uni-directional have to be installed with the seal lip facing the pressure chamber to be effective and it should be ensured that service personnel are aware of this.

Seals are classified as either *piston seals* or *rod seals*. Piston seals are located in a groove cut on the piston, moves along with the piston and bears against the cylinder walls. Rod seals are usually stationery as they are located in a groove cut in the cylinder or in a housing and bears against the piston rod that is free to reciprocate.

In case of piston seals it is the cylinder surface that needs to be honed or ground and super-finished, to prevent rapid seal wear out. In case of rod seals it is the piston/piston rod as the case may be that needs to be ground and super finished. U seals having symmetrical cross section can be used either as rod seal or as piston seal.

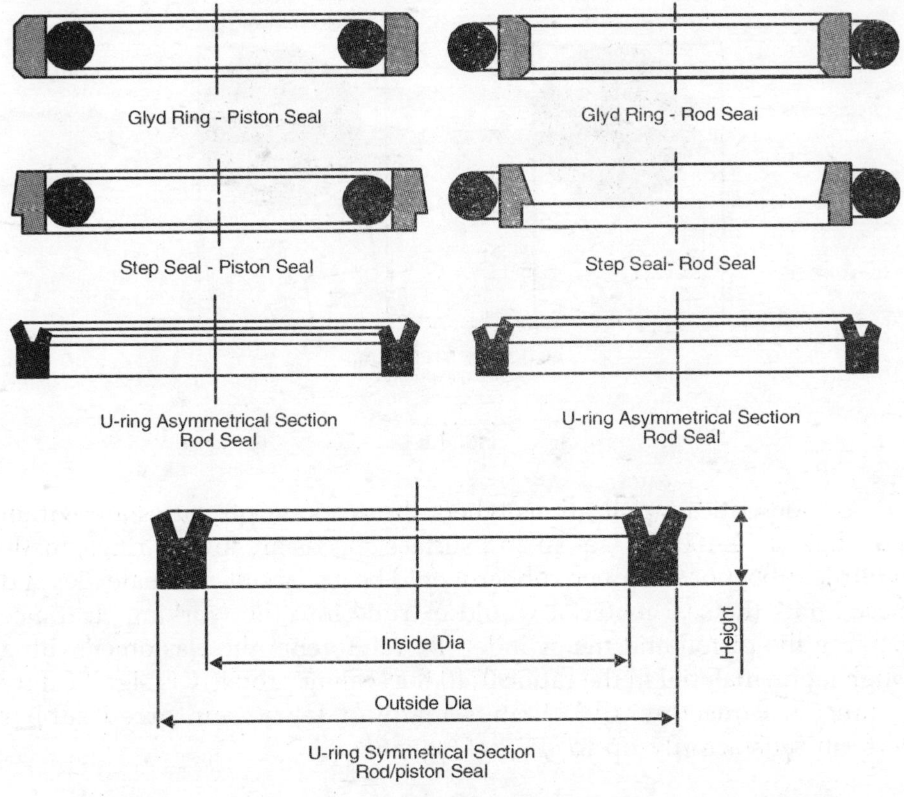

Fig. 7.5

Both step seals and glyd-rings have a circumferential band of Teflon placed either on the inside or the outside diameter of a rubber *O* ring depending upon whether it is a piston seal or a rod seal. The *O*-ring acts as the energiser. These seals are very good for dynamic sealing applications, as they can with stand high reciprocating velocities, elevated temperatures up to 200ºC and pressures up to 800-kgf/sq-cm.

U-seals are a single piece molded construction performing simultaneously as an energizer and as a bearing for the rod. Such a construction can at best be a compromise on both the functions. In step seals and glyd-rings these functions have been separated with two different materials. The rubber *O*-ring performing as an energizer, and the reinforced PTFE compound providing the bearing function. Here step seal needs a special mention due to its unique mechanism of action and its hydrodynamic feature of returning the remaining oil film back to the cylinder when the rod retracts thus literally wiping the rod clean of oil.

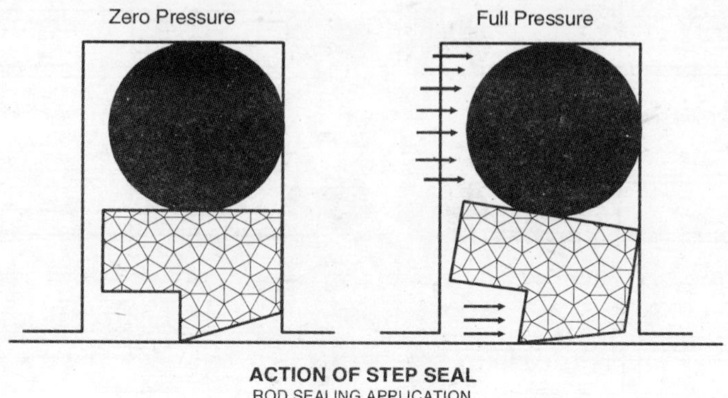

Fig. 7.6

For non-critical applications where the working pressures are within the range of 70–100 kgf/sq-cm and surface speeds are low, U rings, made of nitrile rubber or neoprene rubber would be just about adequate. Beyond these limits the seal material would extrude into the working clearance between the piston and the cylinder. By reinforcing the elastomer with a tough fabric material in the ratio 90 : 10 the sealing property is significantly enhanced. Consequently, U rings made of fabric-reinforced rubber perform satisfactorily up to 175 kgf/sq-cm.

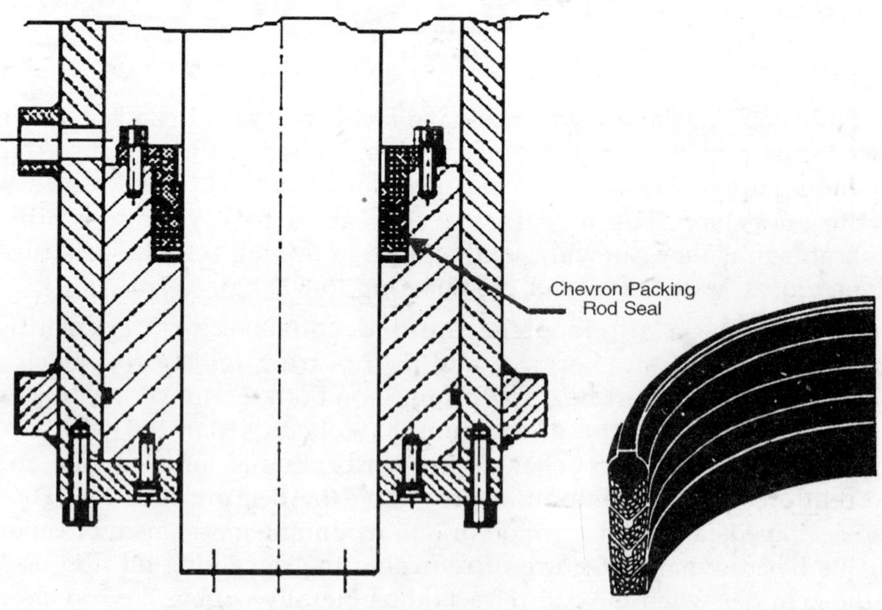

Fig. 7.7

The standard chevron/V pack consists of a male and a female adapter also called the base and the header ring with three or more intermediate rings, thus providing a multi lip sealing. The female adapter acts as the support for the entire pack and is usually made of fabric-reinforced rubber.

In high-pressure applications to prevent extrusion of the female adapter, they are made of alternative materials such as phenolic-fabric, glass-filled nylon or acetal resin. The intermediate rings are interchangeable rubber V- rings, which will prevent low-pressure leakages. The header ring is the primary seal. It is a bonded construction with nitrile/neoprene rubber over rubberized fabric.

The height of the pack can be increased by additional intermediate V-rings. By knowing the effective depth of a single V-ring, the alternative heights can be calculated. Chevron packing are good for pressures up to 500 kgf/sq-cm. They can also withstand very high reciprocating speeds, up to 60 meters per minute. It is, however, important that they should be installed carefully and in well designed grooves. Also in gland-sensitive design, excessive tightening of the gland nut will cause the first ring to wipe the wall completely dry, which makes the succeeding rings wear out, faster. Therefore some small leakage must be allowed to provide a lubrication film between the metal and the rubber. For zero leakage a polyurethane 'V' seal should be included as the outer member of the 'V' stack.

Solid 'V'-packing provides better sealing than split 'V'-packing. If split rings must be used for ease of installation during replacement, the joints should be staggered on successive rings. The function of male and female adapters is to support the rings. The female adapter is more critical and should have the same included angle as the male. V packing/chevron packing are more often used as rod seals than as piston seals. This may be because they require a larger installation space, which is not available on smaller pistons. If installed on large pistons replacement of damaged seals could present problems.

For pressures beyond 175 kgf/sq-cm and in applications where high abrasion resistance is called for and where operating temperatures exceed 50C U-seals made of the thermo-plastic polyurethane elastomer known commercially as Ecopur is the recommended seal. Its outstanding features are, unusually high resistance to abrasion, low compression set, high rigidity and tear strength.

7.3 FACTORS INFLUENCING EFFECTIVE SEALING AND LONG SEAL LIFE

Seal groove width and depth must be maintained exactly to sizes recommended by the manufacturer of the seal being used. If the seal groove is cut on the rod it is easy to measure and maintain the depth. If the groove is cut on the housing, proper inspection tools must be used to inspect the groove depth. Sharp corners of the groove must be blunted to form a round edge, the radius not exceeding 0.5 R This will prevent damage to the seals both during installation and working. Fifteen degree lead-in chamfers of sufficient length which depends on the size of the seal must be provided to prevent damage to the seals during entry. Proper installation tools must be employed where required to prevent damage to the seals during installation or the groove itself must be so designed as to facilitate easy installation.

Concentricity of the seal groove with respect to the rod or the housing is very important. In case of rod seals the piston rod must be ground, hard chromium-plated, and reground so as to offer a smooth wear resistant surface for the seal in contact. In case of piston seals, the cylinder must be ground and honed. The surface roughness in either case should be ≤ Ra 0.4 mm for rubber and polyurethane material and for PTFE based material it should be ≤ 0.2 mm. Less than optimum surface finish with high peaks results in early failure of the seal due to abrasive wear.

Solid contaminants in the system such as grinding dust, mould sand, scale, welding beads etc. contribute to early failure of seals and to re-machining or replacing of damaged mating surfaces.

7.4 OIL RESERVOIR

It is basically a leak proof enclosure fabricated out of steel plates and designed to hold a certain minimum quantity of hydraulic oil with facilities for mounting of hydraulic power unit, consisting of the electric motor, coupling, bell housing and the pump. Other "built in" facilities are, side openings on either side, of sufficient size to enable complete cleaning of the tank, provision for filler cum breather units, drain plug, oil level indicator. The side openings would have steel covers sealed with gaskets and secured by sufficient number of peripheral screws.

Oil tanks usually are an integral part of the machine structure formed by taking advantage of a cavity in the machine base.

This concept is popular because it saves space, is economical and improves the cosmetic appearance. The design of integral oil tank is however more critical, as there cannot be a compromise on the desired volume of the tank or its accessibility to servicing and amiability to routing of hydraulic fluid lines to and from the actuators.

Fig. 7.8. *Integral power pack.*

The other option is a separate oil tank, independent of the machine, which is a self-contained unit. This has the electric motor, the pump and all other system components mounted and connected as per the circuit, through the fluid conductors. Tapping points are provided for pressure and return lines. Such a unit is called *the power pack.*

Fig. 7.9. *Independent power pack.*

Although independent power packs occupy a little more space, their advantages are obvious. They are easy to install and service. They can be fabricated within the factory or entirely bought out, to specifications.

Also, one power pack can be made to serve two or more machines if sufficient dwell time is available on each machine.

Whether integral or independent, the oil tank is usually, either rectangular or L-shaped. The Electric motor and the pump unit are either mounted horizontally or vertically. In vertical mounting the pump is usually submerged in oil. Pumps submerged in oil may be difficult to service without draining the oil from the reservoir, but they eliminate the problem of pump cavitations altogether and considerably reduce noise levels. Vertical units with submerged pumps are more compact and will allow a lot more free space on the top surface for mounting other hydraulic components. L-shaped power packs, have a tall narrow rectangular tank with the pump-motor unit mounted on the horizontal platform, which has a common base. A rectangular tank placed above the pump and motor unit in an overhead configuration is another, option. A shut off valve must be incorporated in either case at the inlet line to the pump to enable servicing the pump without draining the oil tank. A typical rectangular oil reservoir is shown in the figure below. Ease of maintenance should be the philosophy governing the design of hydraulic oil reservoirs. After fabrication, the insides of the tank must be thoroughly cleaned to remove all metal particles.

Sand blasting followed by kerosene cleaning is an ideal solution. The entire insides of the tank is first coated with metal primer and then painted with a sealer to reduce rust formation due to condensed moisture. Flow lines from and to the reservoir such as, suction, return and pressure lines must be provided with sealed slip joint flanges to prevent dirt and dust. Access covers at convenient locations on the tank should be provided to enable removal of suction strainers for cleaning, at regular intervals.

Baffle plate, placed lengthwise in the center of the tank is intended to isolate the suction side from the return side to ensure that the same fluid is not pumped out. Suction and return lines must terminate well below the oil level. Return line ends must be angled to divert the flow towards the tank walls. A Filler-breather must be provided on tank top. The filler-breather should have at least a 75-micron mesh with sufficient airflow capabilities consistent with pump discharge. Installation of a diffuser in the tank return line below the oil level may be required in certain cases to reduce return line velocities and consequent, turbulence, foaming and noise. Since the oil tank is a completely closed unit, oil level indicators are required to indicate the minimum quantity of oil that should always be available in the reservoir for proper functioning of the machine.

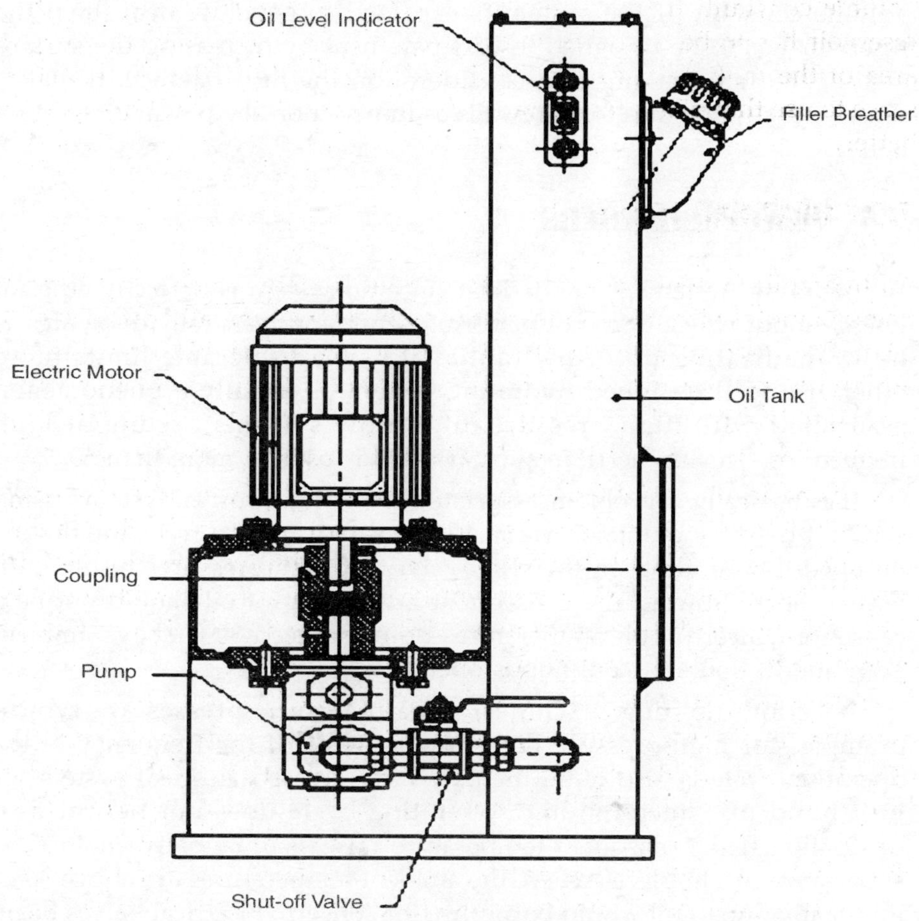

Fig. 7.10. *L-Type power pack.*

These are also available with oil temperature indicating facility. These are inexpensive and perform important tasks. The minimum oil level in the reservoir must be well above the pump suction port in case of submerged pumps or well above the level of the suction strainer, in case of pumps mounted above the reservoir.

Sizing of oil reservoirs is based on the quantity of hydraulic oil that needs to be stored. This depends on the pump delivery and the amount of heat the oil reservoir is expected to dissipate. The thumb rule is, oil quantity required in liters = 3–5 times the pump discharge in LPM. If for instance, a pump discharges 50 LPM, the minimum oil quantity required would be 150 litres or 150,000 cc. An oil reservoir of length 75 cm, breadth 40 cm, and 50 cm wide, would adequately meet the requirement. These dimensions of course, can be altered to suit a specific size keeping the volume constant. If the system is expected to generate heat, then the reservoir has to be designed to dissipate heat by increasing the surface area of the tank, taking into consideration the heat balance. It makes sense to use thin plates for sidewalls as thin sections help heat dissipation better.

7.5 HEAT EXCHANGERS

All hydraulic systems generate heat since no system is perfectly efficient Energy input which does not perform useful work generally dissipates as heat. The heating up of hydraulic oil beyond tolerable limits in an otherwise well designed hydraulic system, is usually a phenomenon associated with high pressure, high flow systems cycling at high frequencies. The input HP in such systems is usually more than 50.

It is basically a problem associated with fluid power systems using constant delivery pumps in which, for most part of the cycle, the fluid is dumped to tank through the relief valves, at high pressures resulting in wasted horse power. The consequent heating up of oil should not be a cause for concern if oil temperatures do not exceed 30ºC above ambient even after 8 hours of continuous operation.

Injection molding machines, huge hydraulic presses are typical examples for high-pressure high flow systems. If the frequency of the operating cycle is also high, then we have created an ideal system for heat build up, since the high operating cycle does not permit heat dissipation. The consequent temperature raise shall be between 40–60ºC above ambient. In places where the ambient temperatures are above 30ºC, this heating up of oil would be a cause for concern. Electrical valves begin to malfunction, leakage past the seals and spool valves due to viscosity changes, would affect the system performance. Poorly designed hydraulic system is another cause for heating up of oil. The flow ratings of all valves and conductors must be adequate to handle the pump flow. In a double acting cylinder with a large piston area to annular area ratio, valves must be selected to handle the discharge out of the rod end of the cylinder and not based on the flow in-to the cap end. Heat exchangers are designed primarily, to overcome this problem. If installed at an appropriate location, which is the tank return line, heat exchangers will help the system to dissipate heat and maintain oil temperatures with in normal limits.

The use of a heat exchanger may also be justified where exacting oil temperature has to be maintained to stabilize oil viscosity.

Basically, there are two types of heat exchangers. *Liquid-to-Liquid and liquid-to-Air type.* Liquid-to-liquid type, are of shell and tube construction consisting of a bundle of small tubes held inside a shell. The coolant flows through the small tubes while the hydraulic oil passes around and between these tubes. The tubes can be either of the straight type or of the U-type. Straight type units have higher thermal efficiency and are less expensive. U-types are best suited for high temperature and high-pressure applications.

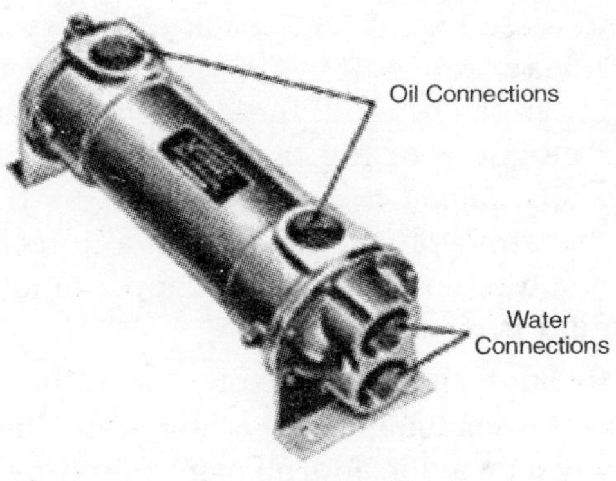

Fig. 7.11. *Liquid-to-liquid heat exchanger.*

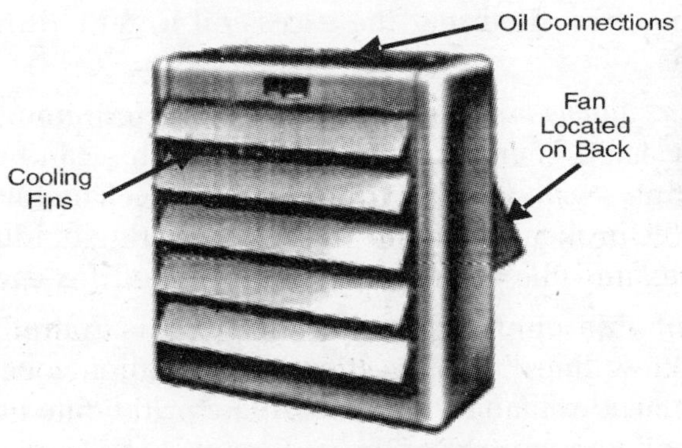

Fig. 7.12. *Liquid to heat exchanger.*

They can accommodate thermal expansion but the tube banks are difficult to clean. The heat exchangers can be of single pass or double pass and can be of parallel flow or counter flow type. The double pass counter flow type provide the maximum heat transfer for a given size.

For hydraulic applications the commonly used coolant is water. Standard liquid-to-liquid units can handle pressures up to 15 kgf/sq-cm. And temperatures up to 150ºC although the actual temperature difference between oil and water should not exceed 90ºC. The mechanism of action of liquid-to-liquid heat exchangers is one of evaporation and condensation. The tubes are basically, an enclosure containing a fluid that can be vaporized and a material that provides capillary action. In what is basically an isothermal process, heat at the input end causes the fluid to evaporate. Vapour travels through the tube to the condenser or out put end. Here the vapour condenses giving up its latent heat and returns to the fluid state and travels back to the system through the evaporator end. Clean and soft water should be used to prevent corrosion and scaling in the tubes. If the water is hard and has excessive salt, it can lead to clogging of the insides of the narrow tube due to scale and dirt deposits. *Liquid to air heat exchangers* transfer heat from hydraulic fluid to the atmosphere, just like an automobile radiator. The air passes over finned tubes made of either copper or aluminum in which the hot hydraulic fluid is circulating.

Air is moved through the core by force or induced draft fans. The liquid-to-air type heat exchangers are portable units that can be used on site. They are however heavier and noisier compared to the liquid-to-liquid type.

Heat exchangers are available as of the shelf components to various sizes and configurations. They are available with temperature control valves water strainers and Bypass check valves. By pass valves protect the heat exchanger by diverting the excess oil to tank during surge and peak pressures.

Selection of a suitable heat exchanger involves determining the total heat equivalent of wasted energy or the heat load. This can be assessed, by analyzing the duty cycle of the hydraulic machine. The total time period of a cycle must be broken up into its phases consisting of idling, approach, work, return etc. and the wasted energy calculated for each phase.

The sum of this energy less the energy dissipated through the reservoirs and flow lines by convection and radiation constitutes the net energy that the heat exchanger must dissipate or the rate of heat transfer. The maximum oil temperature at the inlet point to the heat exchanger can now be determined by using the equation,

$$H = m\,C_p\,(T_I - T_0) \tag{1}$$

Here, H is the heat transfer rate in kilo Joules/sec, C_p is the specific heat at constant pressure which for hydraulic oil = 1.97 kJ/kg °K. T_I is the oil temperature at the inlet to the heat exchanger, which typically lies between 55 – 65ºC. T_0 is the oil temperature at the outlet. This is the desired oil temperature that the heat exchanger must maintain. A suitable value may be assumed for T_0 consistent with the system requirement. m is the mass flow rate determined by the equation,

$$m = \rho \times q \text{ kg/sec} \quad (2)$$

here, ρ is the density of hydraulic oil in kg/m^3 and q is the oil flow rate in m^3/sec.

If the difference between T_I and T_0 is large then that will justify the need for a heat exchanger. If the difference is only marginal the heat exchanger may be dispensed with. By sizing the surface area of the reservoir, so that it can transfer enough heat from oil to the atmosphere by conduction, convection and radiation process the net wasted energy that the heat exchanger must dissipate can be reduced, thus reducing the size of the heat exchanger.

After ascertaining the need for the heat exchanger, the size of the heat exchanger, in terms of the quantity of water required to dissipate the heat and maintain the desired temperature, must be determined based on the equation,

$$Q = (860)\,(P_w)/T_I - T_w \quad (3)$$

where P_w is the wasted energy in kW, T_I is the maximum temperature of the oil at the inlet to the heat exchanger, in ºC, T_w is the water inlet temperature in ºC. Q is the water flow rate through the heat exchanger in liters/minute. When the flow capacity as determined from equation (3), is much larger than the capacity of available heat exchangers, then two smaller heat exchangers of equal size can be used in parallel. Standard heat exchangers would have oil to water ratio of either 1 : 1 or 2 : 1.

This means that for every liter of oil circulated one or half a liter of water must be circulated, to achieve the desired oil temperature. The size and therefore the cost of a heat exchanger, depends largely on the initial temperature difference between the inlet oil and water. If this difference decreases the cost decreases. For effective cooling inlet water temperatures should not exceed 20–25ºC. The pressure drop that the hydraulic system can permit is another factor that has a direct bearing on the size and cost of the heat exchanger. Ideal allowable pressure drops should be between 3.5 to 5 kgf/sq-cm.

If it becomes necessary to install a heat exchanger in an existing system, then the wasted energy may safely be assumed equal to 20 percent of the connected power. The value of T_I may be obtained directly by measuring the maximum temperature of the oil in the tank and the heat exchanger sized, based on Eq. (3) above. Shell and tube heat exchangers can be mounted either vertically or horizontally.

Problem. *A Hydraulic machine has the following duty cycle.*

Idle at 15 kgf/sq-cm for 30 seconds. Clamp work piece at 100 kgf/sq-cm for 5 sec.

Approach at 15 kgf/sq-cm for 2 seconds. Perform work at 300 kgf/sq-cm for 3 seconds. De-clamp. Return at 15 kgf/sq-cm for 2 seconds. The pump flow is 100 LPM. Total surface area of the oil reservoir is 2.5 sq metres. Hydraulic pipeline is ϕ*25 mm × 2500 mm long. Calculate the net wasted energy that needs to be dissipated and recommend a suitable heat exchanger if necessary. Assume the following values. Ambient temperature = 20ºC. Volumetric efficiency of the pump = 0.85. Density of oil = 860 kg/m*^{3}*. 1 kWh = 3.6 ×* 10^6 *Joules.*

Solution. Power dissipated in kW is given by the equation, $PQ/600 \times \eta$, where, P is the Pressure in kgf/sq-cm, Q is the flow in Liters/mt., η is the volumetric efficiency of the pump.

Power dissipated during the idle cycle while passing oil through valves and pipe lines

$$= 15 \times 100/600 \times 0.85 = 2.94 \text{ kW}$$

Since this phase lasts for 30 seconds the energy consumed

$$= 2.94 \times 30 / 3600$$

$$= 0.0245 \text{ kWh or } 0.0245 \times (3.6 \times 10^6)$$

$$= \mathbf{8.82 \times 10^4 \text{ Joules}}$$

Power dissipated during clamping

$$= 100 \times 100 / 600 \times 0.85 = 19.6 \text{ kW.}$$

Since this phase lasts for 5 seconds, the energy consumed

$$= 19.6 \times 5/3600 = 0.027 \text{ kWh or}$$

$$0.027 \times (3.6 \times 10^6)$$

$$= \mathbf{9.72 \times 10^4 \text{ Joules}}$$

Power dissipated during approach

$$= 100 \times 15/600 \times 0.85 = 2.94 \text{ kW.}$$

Since this phase lasts for 2 seconds, the energy consumed

$$= 2.94 \times 2/3600 = .001 \text{ kWh or}$$

$$0.001 \times (3.6 \times 10^6)$$

$$= \mathbf{3.6 \times 10^3 \text{ Joules}}$$

Power dissipated during work phase

$$= 100 \times 300 /600 \times 0.85 = 58.8 \text{ kW.}$$

Since this phase lasts for 3 seconds, the energy consumed

$$= 58.8 \times 3/3600 = 0.049 \text{ kWh or}$$

$$0.049 \times (3.6 \times 10^6)$$

$= \mathbf{17.64 \times 10^4}$ **Joules**

Power dissipated during return

$= 100 \times 15/600 \times 0.85 = 2.94$ kW.

Since this phase lasts for 2 seconds, the energy consumed

$= 2.94 \times 2/3600 = 0.001$ kWh

or $0.001 \times (3.6 \times 10^6)$

$= \mathbf{3.6 \times 10^3}$ **Joules**

Total power dissipated in 42 seconds = $\mathbf{2.1 \times 10^5}$ **Joules**. In terms of average wasted power this will be, $2.1 \times 10^5/3.6 \times 10^6 = 0.058$ kWh = 0.058 $\times$ 3600/42 =5 kW. Available hydraulic power = $300 \times 100/600 \times 0.85 = 58.8$ kW. Wasted power is 8.5 % of available power.

If we re-arrange Eq. (1) we have

$$(T_I - T_0) = H/m \times C_p \quad (4)$$

$H = 2.1 \times 10^5/1000 \times 42 = 5$ kJ/ second, $m = 860 \times 0.001 = 0.86$ kg/sec Substituting the above values in Eq. (4) we have $(T_I - T_0) = 5/\ 0.86 \times 1.97$ = 2.95. Since this difference is negligible there is no need for a heat exchanger.

Heat dissipated by the oil reservoir and hydraulic pipelines can be calculated as follows.

Heat dissipated by the oil reservoir = $\Delta T \times A \times K/3600$ kW,

where, ΔT is the difference between the maximum oil temperature and the ambient temperature in ºC, A is the surface area of the reservoir in m^2, and K = average rate of convection from reservoir = 6 to 10 kJ/cm^2 hr. depending upon the condition of the surroundings. For poor air circulation assume a value = 6. The surface area of hydraulic pipelines is given by pdl, where d = outside diameter of the pipe and l = length in appropriate units.

8

APPLICATION ENGINEERING

8.1 PREVIEW

Having obtained a basic knowledge of the principles of hydraulics and an insight in to the functions of various hydraulic components, their relative merits in terms of performance, cost and ready availability, the student is now, in a position to practice application engineering.

Application engineering begins with the preliminary design of the hydraulic circuit. A clear understanding of what is expected of the circuit, the cost constraints, the productivity and degree of sophistication required is important before proceeding further. Hydraulic circuit design, involves choosing the appropriate hydraulic components required to achieve a certain end result, representing them in their graphical form on paper in a certain order, analyze their functions individually and collectively so as to ensure overall system compliance.

Hydraulic circuits are of two types. Open loop circuits (without feed back) and closed loop circuits (with feed back). Open loop circuits can be either constant flow circuits or demand flow circuits. Both constant flow and demand flow circuits can be further classified as either 'open circuits' or closed circuits. In a typical constant flow circuit pump unloads to tank in the valve neutral position and if the pump delivery is admitted to one end of the actuator return oil from the other end is directed to tank or is fed back to the pump depending upon whether it is an open circuit Fig. 8.1(a) or a closed circuit Fig. 8.1(b). Generally constant flow circuits are the most economical provided they meet circuit requirements. A typical demand flow open circuit Fig. 8.1(c) uses a fixed displacement pump, an unloading valve with or without an accumulator. In a demand flow circuit, all ports are blocked in the valve neutral position. If a variable displacement pressure compensated pump is used then the accumulator and the unloading valve may be dispensed with Fig. 8.1(d). Demand flow closed circuit Fig. 8.1(e) is similar to constant flow closed circuit except that instead of a fixed displacement pump a pressure compensated variable delivery pump is used.

A closed loop system is always a servo-controlled system with feedback but a servo-controlled system need not be a closed loop system. (Refer Figs 8.1(g) and (h) below. In any hydraulic circuitry, the ultimate purpose is to control the velocity, position, direction of motion and the force exerted by the actuator. Force exerted, may be controlled by pressure regulators and pressure reducing valves. Direction of motion, of the actuator may be controlled by the use of directional control valves. For speed control we have the flow control valves and variable delivery pumps. It is the precise control of position that often becomes the problem.

Consistent and reasonably accurate control of the position of the actuator is possible through electronic timer controls or through cam actuated limit switches.

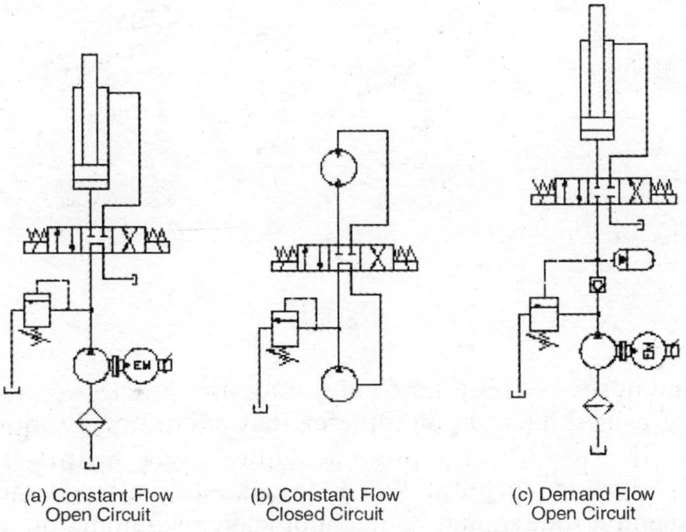

(a) Constant Flow Open Circuit

(b) Constant Flow Closed Circuit

(c) Demand Flow Open Circuit

(Contd.)

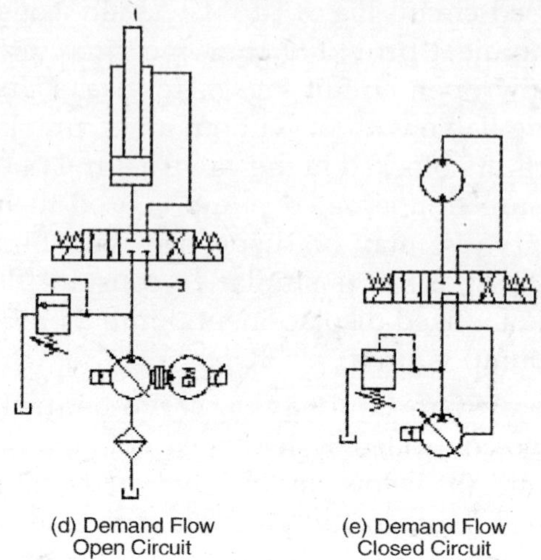

(d) Demand Flow Open Circuit

(e) Demand Flow Closed Circuit

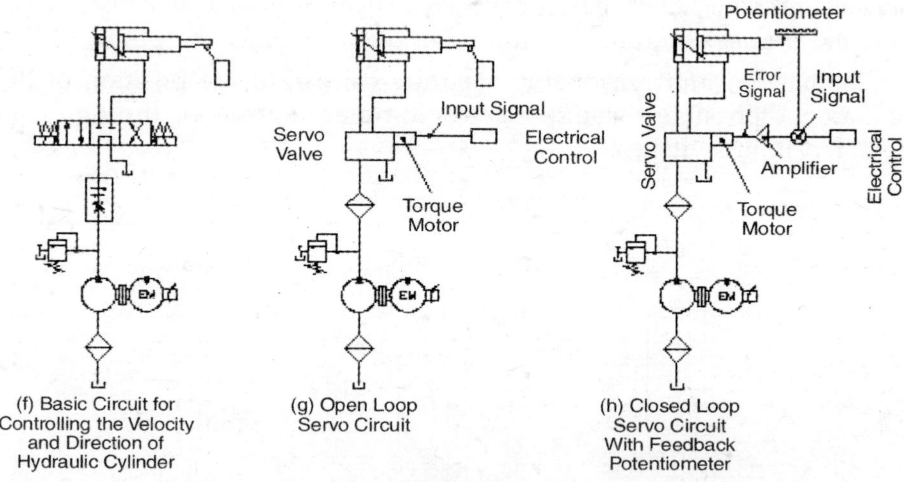

(f) Basic Circuit for Controlling the Velocity and Direction of Hydraulic Cylinder

(g) Open Loop Servo Circuit

(h) Closed Loop Servo Circuit With Feedback Potentiometer

Fig. 8.1

For highly precise position control only the closed loop servo circuit employing a feed back potentiometer that continuously monitors the position of the piston rod, offers the solution. (Fig. 8.1(h)). In a servo-system, the directional control valve is replaced by a servo valve, which has an integral torque motor. A potentiometer, a summing up device and an amplifier in a closed loop control mode would complete the circuit requirement. It may be noticed the closed loop servo control system displaces the flow control valve, the return line check valve and the cam controlled limit switch that is required in a conventional circuit. A servo valve combines the functions of a directional control valve and the flow control valve.

Since a detailed analysis of the servo-controlled system is beyond the scope of this book we shall go no further than this.

8.2 ANALYSIS OF LOADS

The design of any hydraulic circuit must begin by analyzing the kind of load that the actuator is expected to handle. The basic types of loads are, resistive loads where the load reaction opposes the movement of the actuator, over running loads where the load reaction has the same direction as the load, or inertial loads executing linear or rotary motion in a steady state which follows the law $F = ma$ or $T = J\alpha$, when a change of state occurs. Here J refers to the polar moment of inertia and α is the angular acceleration.

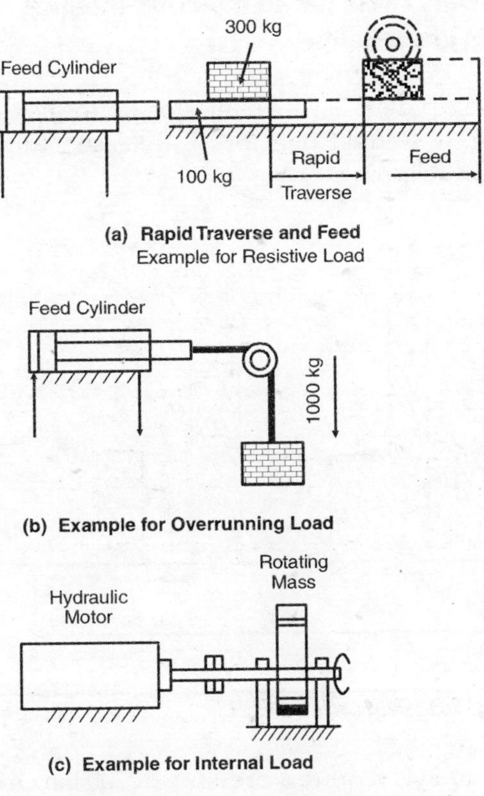

Fig. 8.2. *Examples for resistive load.*

During the start up phase all loads are partly inertial. Therefore loads must be predominantly inertial to be considered as such. Large hydraulic presses, extrusion presses, broaches and burnishing machine rams could be considered as predominantly resistive loads. Machine tool carriages, feed cylinders have a variable resistance load with a strong component of inertial load during acceleration.

After identifying the type of load, the next step is to determine the optimum method to transfer the load. If we decide to accelerate the load to a pre-determined velocity and transfer the load at constant velocity a low volume pump could meet the low constant velocity requirement but has to operate at higher pressures due higher pressure required during acceleration.

On the other hand if we choose to accelerate gradually for half the distance and decelerate gradually towards the other half although a low-pressure pump would meet this requirement a high volume pump would be required since the average velocity is higher. The other alternative is rapid acceleration for most part of the cycle followed by rapid deceleration towards the end and this leads to a low-pressure high-volume pump.

8.2.1 A Typical Duty Cycle for an Injection-molding Machine is given Below

1. Close dies. 2. Clamp dies. 3. Move injector up to dies. 4. Inject. 5. Dwell period for cooling. 6. Retract injector. 7. Release clamps. 8. Open dies. 9. Dwell period for stripping.

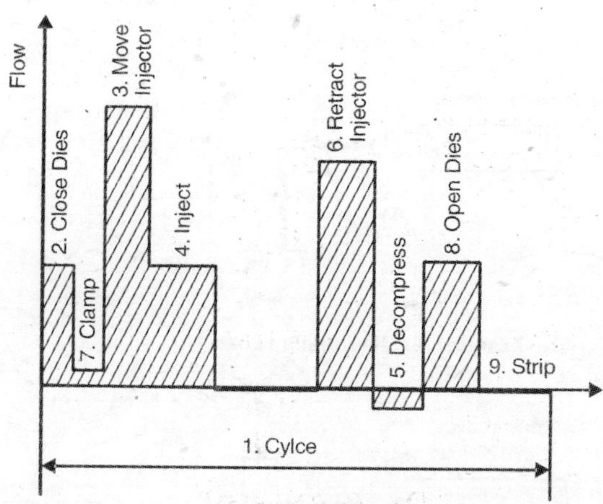

Fig. 8.3. *Duty cycle for injection-moulding machine.*

Operations (2) and (4) require a pressure of 200 bar. All other hydraulic motions can be achieved with a pressure of 80 bar. The flow required during each operation is shown is shown in Fig. 8.4.

Although pressure required during operations (2) and (4) is high, the flow required during this period is very small. Therefore the pump may be selected from considerations of maximum flow required while limiting its pressure requirements to the over all system requirement which is 80 bar.

The 200 bar required for operations (2) and (4) could be derived from the 80 bar supply by using an intensifier in the circuit. Alternatively, an accumulator can also supply the high-pressure flows. The accumulator could be charged by a separate smaller capacity, high-pressure pump running continuously, relieving through unloading valves for short periods when the accumulator is fully charged.

Since the flow requirements for most part of the cycle is not constant a variable delivery pump capable of meeting the largest flow required with a continuous rating of 80 bar may be the ideal choice.

This pump could as well be a pressure compensated pump. The swash plate angle would automatically adjust itself to supply the flow demanded by the system throughout the cycle. More accurate analysis could be made by graphically representing the load magnitude cycle, pressure pattern cycle and power demand pattern cycle.

A hydraulic circuit may be drawn using block diagrams or cut away sections of hydraulic components or using graphical symbols as approved by IS: 7513–1974 or ISO standards. Hydraulic circuit diagrams drawn using graphical symbols are the most popular. They are easy to draw, enable the reader to comprehend the functions more easily and since there is no particular language involved has worldwide acceptance. Graphical symbols drawn as per ISO standards for various hydraulic components have been listed elsewhere in this book. Circuit diagram using cut away sections of hydraulic components is shown in circuit-5 below.

8.3 CIRCUITS USING HAND PUMP

1. Workshop Press. These presses are used in industry and general engineering workshops for day to day assembly and disassembly of press fitted parts, bearings and for such odd jobs like bend removal of shafts, plates etc. These are not intended for high volume production. These machines develop between 5–50 tonnes.

The circuit shown is for a 7.5 tonne press. A two-stage hand pump was chosen.

This pump has 'built-in' check valves, relief valves and pressure release valves. The pump chosen can deliver 50 cm^3 per stroke up to 30 bar and can automatically change over to 3 cm^3 per stroke beyond this pressure

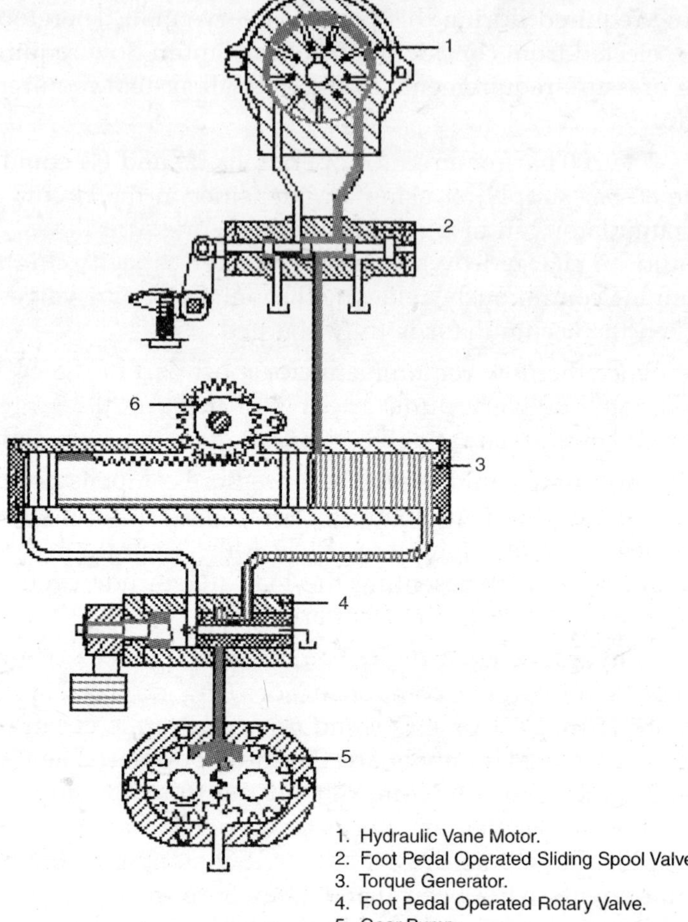

Fig. 8.5

range. The pump chosen has a maximum capacity of 350 bar. The capability to deliver higher volumes at lower pressures makes it possible to set a faster approach speed for the press. The piston has a diameter = 6.3 cm and therefore an area = 31.17 cm^2. In this case the piston would move @ 50/31.17 = 1.6 cm per stroke during fast approach and theoretically develops the designed tonnage at a working pressure of 240 bar. The net force developed will be the theoretical force less the force required to over come the spring stiffness, which indeed is a very small percentage of the total force. The piston rod of the single acting spring returned cylinder would return to its original position upon release of pressure with the help of hand knob provided on the pump. The return speed would depend on the flow passage area available for the exhaust oil to return to tank.

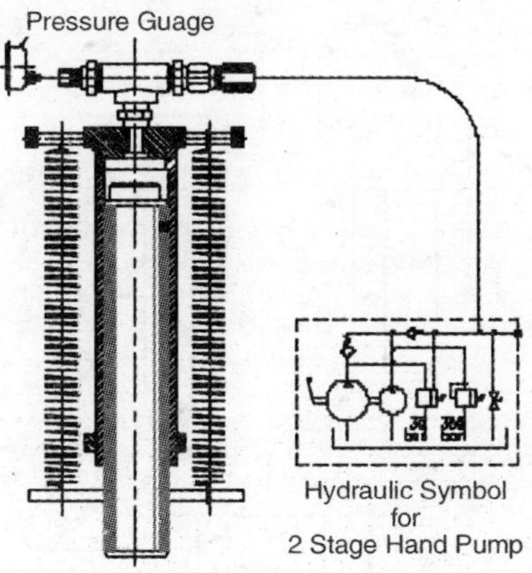

Fig. 8.6

8.4 CIRCUITS USING CONSTANT DELIVERY PUMPS

1. 60 Tonne Press Brake. This machine was designed to bend 4 mm, mild-steel plates up to 1 metre in length using two double acting cylinders. Each cylinder bore has an area of about 176 cm^2 and the required tonnage is obtained by applying a pressure of 175 bar using a gear pump which is rated for 200 bar.

Gear pump (1) was chosen to keep the costs down since the production was intermittent. A pump delivering 10 lpm or 166 cm^3 per second at 1440 rpm of the electric motor (2) was chosen as this would give an approach speed of $166/176 \times 2 = 0.45$ cm/sec or $\approx$ 4.5 mm/sec. Since the die gap was only 50 mm including the distance the tool would travel after contacting the die, this delivery was considered sufficient, as the job would be accomplished in, theoretically, five seconds. Based on the pump delivery and working pressure a 5 HP electric motor was found adequate.

An over all efficiency of 75% was considered for the gear pump while calculating the HP required. The direct operated pressure relief valve (4) was locked at 175 bar. The hand operated four-way three position, spring centered directional control valve (4) with pressure port open to tank in the valve neutral position and cylinder ports blocked was chosen since, it was required that the bending tool must be held in position against gravity during the neutral position. Motion of hand lever 'forwards' and 'backwards' would accordingly put the piston rod and hence the bending tool in the 'approach' or 'return modes. A 40-litre oil tank was found adequate based on pump delivery.

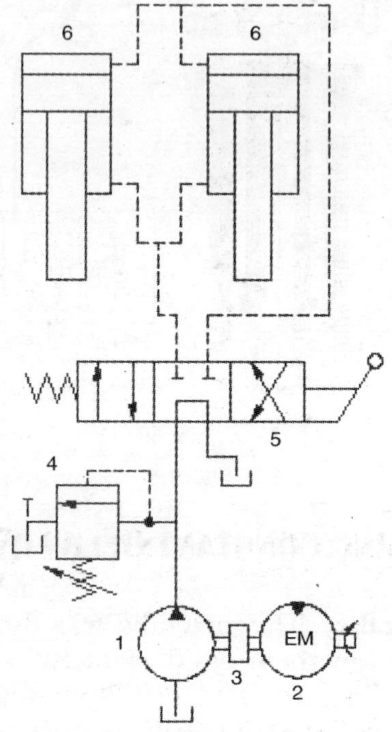

Fig. 8.7

2. Die-cutting Machine. This machine is used to cut various profiles on materials such as leather, Rubber, Plastic, Card-board etc., using profiled steel knives as cutting dies. These machines are designed to develop up to 20 tonnes between the arm and the table. The arm is fixed to the column that is guided in a housing. The inside diameter of the column acts as the cylinder in which the piston and the piston rod is free to slide.

In this particular design, the piston and the piston rod are fixed and therefore are stationary. The cylinder and the column together are free to move 'up and down' instead.

Pump (3) supplies oil to the cap end of the cylinder through drilled holes in the piston rod as soon as the electric motor (1) is started. The column therefore starts moving up and stops as soon as it registers with a hole in piston rod that diverts the pump supply back to tank. The opening and closing of this hole can be regulated, by rotating the hand wheel located at the top of the column. The wheel is connected through linkages to a plunger that slides up and down the piston rod on the cap end thus covering and uncovering the tank port. This in turn controls the position of the column with respect to the table and determines the optimum die gap or the 'day-light'.

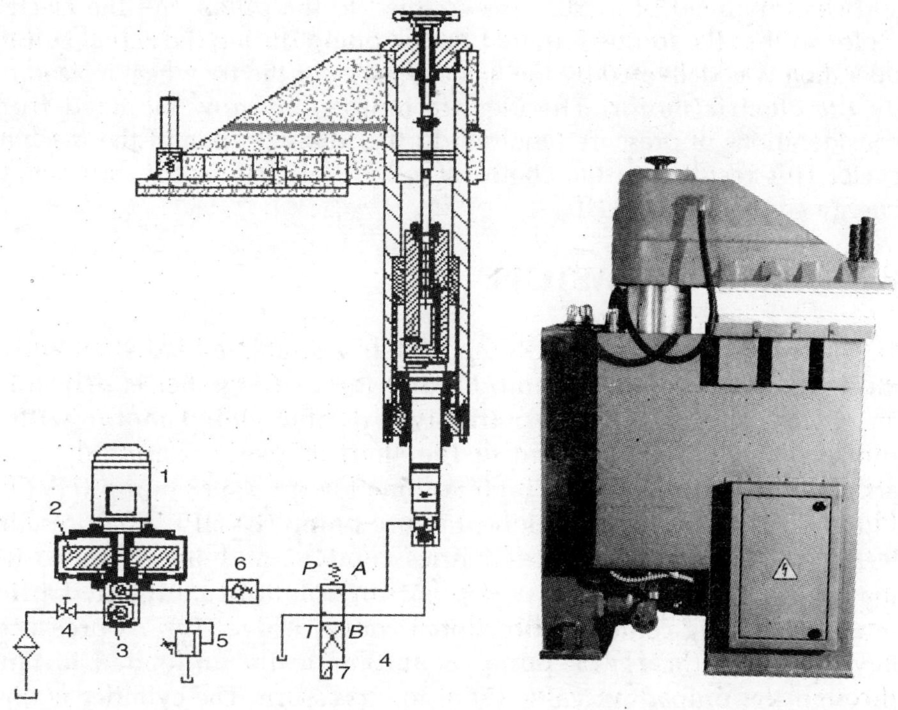

Fig. 8.8

The two position 4 way spring centered solenoid operated directional control valve (7) is energized through push button electrical switches and the spool shifts to divert pump flow to the rod end of the cylinder and the arm starts moving down till it makes contact with the die and applies the force required to cut the material underneath. An electronic timer may be set to count the time required for the arm to move this distance and de-energize the solenoid valve automatically at the end of this time so that the arm returns to its original position. Alternatively, a cam may be set to actuate a limit switch that triggers the valve back to neutral. Considering the high volume of production expected of the machine a 20 lpm discharge rating for the pump at 1440 rpm was considered essential.

The net area where the cutting force is developed is 100 cm^2, which means 20 tonnes required could be obtained only if the pump chosen is capable of exerting at least 200 bar.

A Vane pump was therefore chosen. A pump delivering 20 lpm at 200 bar, even if we assume an overall efficiency of 95 percent for the pump would require at least a 9 HP electric motor. Since the actual cutting operation lasts for not more than a second or two, a fly wheel with sufficient moment of inertia was coupled to the pump and the electric motor so that the torque required by the pump during the actual cutting operation was delivered by the kinetic energy of the fly wheel instead of, by the electric motor. The electric motor may now be sized from considerations of pressure required during idling portion of the machine cycle. This resulted in the choice of a 2 HP electric motor thus saving energy equivalent to 9 HP.

8.5 MULTI-PUMP SYSTEMS

1. 100 Tonne Press Circuit. Pump (2) has a double-ended shaft and is mechanically coupled to pump (3) and the two together is driven by an electric motor (1). Alternatively a double-ended motor with a pump coupled at either end of the shaft is also a commonly used arrangement. Pump (2) is a high volume low-pressure pump (HVLP). Pump (3) is a low volume high-pressure pump (LVHP). At a pressure less than 30 bar the pump deliveries combine and supply oil to the main ram through a 4 way 3-position solenoid controlled pilot operated, spring centered directional control valve (10). At pressures beyond 30bar the HVLP pump is automatically un-loaded to tank through the unloading valve (5) at low pressure. The cylinder is now served by the LVHP pump alone.

The two pumps together supply 55 liters of oil per minute. This enables the ram with an area 500 cm^2 to raise at the rate of 18 mm per second till work resistance is encountered and thereafter the HVLP pump unloads and the LVHP pump delivery of 10 lpm would be sufficient to build up the working pressure which in this case is about 200 bar. In the valve neutral position the combined delivery is diverted to tank through the return line filter (8). The electric motor may be sized based on the LVHP pump delivery and the maximum system working system pressure. Accordingly a 5 HP electric motor was chosen for this application. Energizing solenoid *a* (LH solenoid) of valve (10) will shift the position of the valve spool and this diverts pump flow to the cylinder. This causes the cylinder to move up at a speed controlled by the combined delivery from both the pumps.

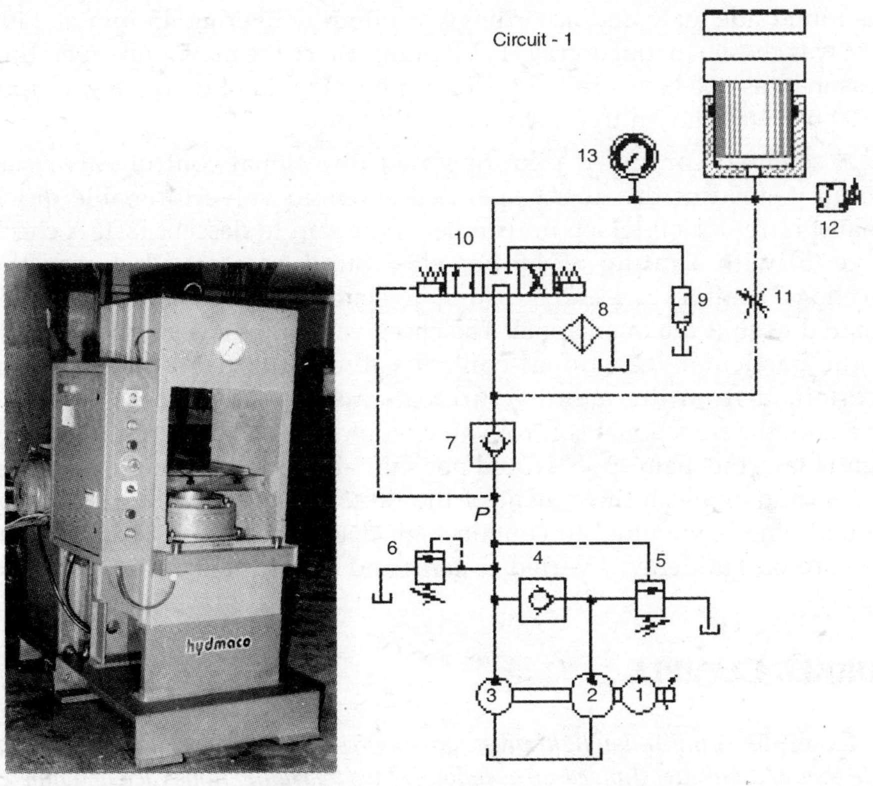

Fig. 8.9. *Tonne electro-hyraulic press and press circuit designed.*

As soon as the work resistance is encountered, the consequent pressure increase causes the HVLP pump to unload through the unloading valve (5). The LVHP pump will now pressurize the ram and this Pressure, is held for the duration set by the timer, not shown in the circuit which receives its signal from the pressure switch (12). After the timer times out, it signals solenoid *a* of valve (10) to return to neutral. The valve is held in this position for one or two seconds to give sufficient time for the cylinder to decompress through the needle valve (11). Solenoid *b* (RH solenoid) is now engaged to bring the ram back to its original position thus completing a cycle.

The above circuit is generally called the 'High-Low' circuit. Functionally it is a HP limiting circuit and technically, it is a load sensing circuit since the pump delivery adopts itself to load requirements.

The unloading pressure of 30 bar and the maximum system pressure of 200 bar can both be set/varied by adjusting the valves (5) and (6) respectively. Since the unloading pressure was only 30 bar a gear pump was found adequate and accordingly a pump delivering 45 lpm at 1400 RPM was chosen to function as HVLP pump. Since the maximum operating pressure was 200 bar, a radial piston pump capable of delivering 10 lpm at 300 bar was chosen to function as LVHP pump.

A solenoid controlled pilot operated directional control valve was chosen instead of the direct operated solenoid valve to enable quick exhaust of return oil which in turn helps the ram to descent fast. A check valve (5) with a rating of 4.5 bar was found necessary between the directional control valve and the pump to generate sufficient pilot pressure required to shift the main spool. The check valve pressure rating depends on the particular directional control valve being used and must be ascertained from the manufacturer. If this pressure is less than the optimum, the directional control valve would not shift at all. If it is only slightly less, the main spool would not shift all the way through, resulting in slow and sluggish movement of the ram. The diffuser (8) installed in the tank line is intended to contain high discharge velocity of the high-pressure oil suddenly diverted to tank and consequent turbulence and noise.

WORKED EXAMPLE

Example: *A multi-day light press for compressing chipboard has a minimum cycle time of 3 minutes divided up as follows. Two hydraulic rams each of diameter 160 mm are used. Pressure required to overcome seal friction and make up for pressure losses in the pipeline is 0.1 % of the maximum system working pressure. It is required to size each pump and determine its un-loading or relief pressure setting, to size the accumulator and determine a suitable pre-charge pressure and to determine the motor shaft horse power assuming 85% overall efficiency for the pumps.*

1. *Load un-compacted sheets in to the press* — *60 sec*
2. *Close platens, 500 mm stroke* — *15 sec*
   (Load, constant at 500 kgf)
3. *Initial compacting, 100 mm stroke* — *5 sec*
   (Load increasing to 15,000 kgf)
4. *Intermediate compacting, 40 mm stroke* — *5 sec (load 30,000 kgf)*
5. *Final compacting, 12 mm stroke* — *5 sec*
   (Load, increasing to 60,000 kgf)
6. *Open platen, 600 mm stroke* — *25 sec*
7. *Strip compacted sheets from press* — *60 sec*

4 positive displacement pumps coupled to a single-electric motor are used.

Pump (1) is used to charge a Hydro-pneumatic accumulator that is used to close the platen. Each time the accumulator is charged the pump should automatically unload. Pumps 2, 3, and 4, are used in combination to achieve three compacting speeds, pumps 2 and 3 being unloaded after initial and intermediate compacting, in that order.

Solution. Area of each ram = 16^2 (0.7854) = 201 cm^2. Area of 2 rams = 402 cm^2. Since the maximum load is 60,000 kgf, the maximum operating pressure would be 60,000/ 402 = 150 bar. Since the pressure required to over come seal friction and pipeline losses is 0.1 % of 150 bar, the working pressure would be 150 + (150 × 0.1) = 165 bar.

Platen closing speed = 500 mm/15 sec = 33.33 mm/sec or 3.33 cm/sec. Rate of fluid required to attain this speed = 2 (201 × 3.33) = 1338 cm^3/sec or 80319.6 cm^3/mt or say 80 litres/minute or 1.33 litres/sec. Since, all of this flow must be supplied by the accumulator in 15 sec. the volume of oil the accumulator must supply will be (1.33 × 15 sec) = 19.95 or say 20 litres (Δv).

It is assumed that the accumulator will supply 20 litres in 15 seconds as the pressure decreases from 75 bar to 25 bar. Maximum accumulator pressure P_2 = 75 bar. The gas pre-charge pressure at room temperature $P_0 = 0.25(P_2) = 0.25 \times 75 = 18.75$ or say 20 bar. Minimum operating pressure $P_1 = 1.1(P_0) = 1.1\ (20) = 22$ bar.

It is assumed that charging and recharging takes place slowly so that the process is adiabatic.

Accumulator size V_0 required

$$= \Delta v / \{(P_0 / P_1)^{1/1.4} - (P_0 / P_2)^{1/1.4}\} \qquad (1)$$

Substituting, the relevant values in Eq. (1) above, the accumulator size $V_0 = 20/\{(20/22)^{1/1.4} - (20/75)^{1/1.4}\}$ = 36.75 or say 40 litres. The minimum operating pressure P_1= 22 bar is more than adequate to hold the weight of the ram and the platens, against gravity.

The charging of accumulator must take place with in the time frame of 145 seconds available during operations 1, 6 and 7 when pumps (3), (2), and (1) are idling. Assuming the charging will be complete in100 seconds or1.70 minutes, pump delivery required will be 40/1.70 = say 24 lpm. Since the maximum operating pressure is 75 bar, the input HP of the electric motor required to drive pump (4) = (75 × 24) /600 × 0.746 × 0.85 = 4.7 say, 5 HP.

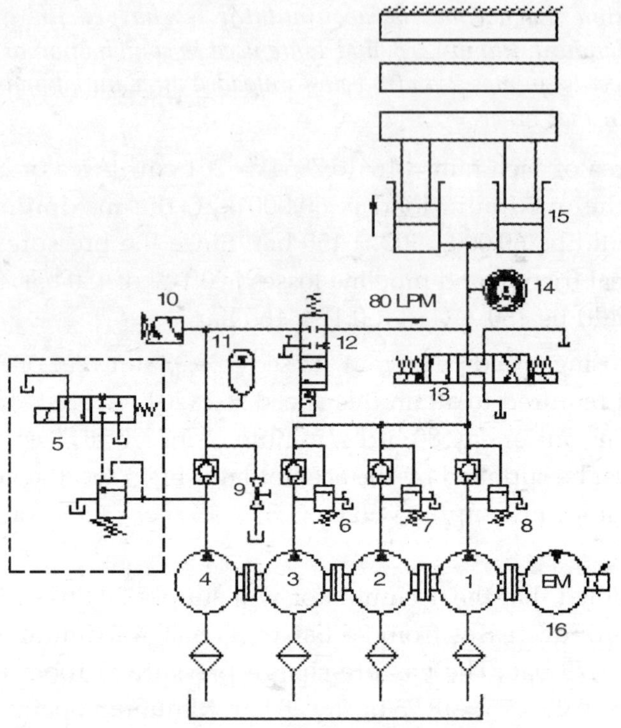

Fig. 8.10. *Hydraulic circuit for multidaylight press.*

The dual function pressure switch (10) signals the solenoid of the solenoid controlled relief valve (5) to shift so that pump (4) can be relieved to tank at low pressure as soon as the accumulator is charged to 75 bar. The spool of valve (5) would remain shifted till the compacting operations are complete and the platen is ready to descend. A timer is intended to count this time and de-energize the valve so that the accumulator can be re-charged in readiness for the next cycle.

The input HP of the electric motor required during this phase would be, (165 × 6)/600 × 0.746 × 0.90 = 2.5 or say 3 HP. The relief valve (8) is set to 165 bar.

Pump (2) unloads after intermediate compacting. The compacting pressure during this phase will be (30,00 kgf/402 cm^2) = 74.6 bar or say 75 bar + 15 bar required to overcome seal friction and pipeline losses = 90 bar. The flow required during intermediate compaction will be (402 cm^2 × 4 cm) = 1608 cm^3. This is the volume of oil that must be delivered in 5 sec. The rate of flow required will therefore be, 1608/5 = 321.6 cm^3 /sec or 19.26 lpm or say 20 lpm less 6 lpm available from pump (1) = 14 lpm.

The input HP of the electric motor required during this phase would be, (90 × 20)/ 600 × 0.746 × 0.85 = 4.7 or say 5 HP. Pump (2) unloads at 90 bar after intermediate compacting.

Pump (3) unloads after initial compacting. The compacting pressure required will be 15,000 kgf/ 402 = 37.3 bar + 15 bar required to over come seal friction and other losses = 52.3 bar or say 55 bar. The flow required during initial compaction will be (402 cm^2 × 10 cm) = 4020 cm^3. This is the volume of oil that must be delivered in 5 sec. The rate of flow required will therefore be, 4020/5 = 804 cm^3 /sec or 48, 24 or say 50 lpm less 20 lpm available from pumps (1) and (2). The in-put HP of the electric motor during this phase would be, (55 × 50) /600 × 0.746 × 0.85 = 7.3 say, 7.5 HP. Pump (3) unloads at 55 bar after initial compacting.

Answer

1. 7.5 HP, electric motor would adequately meet the requirement.
2. Solenoid controlled pilot operated relief valve (5) is set to relieve pump (4) at 75 bar. Un-loading valves (6), (7) are set to unload at 55 bar, 90 bar respectively and relief valve (8) is set at 165 bar.
3. The size of the accumulator would be 40 liters with a maximum pressure rating = 100 bar
4. The gas pre-charge pressure required would be 20 bar.

Pumps (4), (3), (2) and (1) should be capable of delivering, 20, 30, 14 and 6 lpm respectively at pressures 75 bar, 55 bar, 90 bar and 165 bar.

The operator starts the electric motor (16), which drives all the four pumps. Pumps (1), (2) and (3) together dump the oil back to tank through the directional control valve (13). Pump (4) during this phase is charging the accumulator (11) and will be relieved to tank through valve (5) at low pressure as soon the set pressure is reached. The operator first energizes the solenoid operated two-position 4-way directional control valve (12), which in turn shifts the spool permitting free flow from the accumulator to the two cylinders resulting in the upward motion of the rams and platen closure at the desired speed of 3.33 cm/sec.

As soon as the accumulator pressure drops to 25 bar the pressure switch (10) de-energizes valve (12) thus permitting the internal spring to shift the spool back to its original position. The ram is now in its top most position. Solenoid valve (13) may now be energized.

The flow from pumps (3), (2) and (1) is now diverted to the two rams and compacting phase begins during which phase pumps (2) and (3) are set to unload automatically after their set pressure is reached whereas pump (1) will be relieved to tank when valve (13) shifts from $P \rightarrow A$ to $P \rightarrow B$. Valve (13) will return to neutral after the platen descends.

8.6 CIRCUIT FOR FAST APPROACH AND SLOW DIE CLOSING

A machine intended for high volume production has a high piston velocity. If not controlled the high-speed platen approaching the job instead of making a smooth contact will 'bang' on the job. This is not desirable. In all such cases 'rapid traverse and feed circuits' are employed.

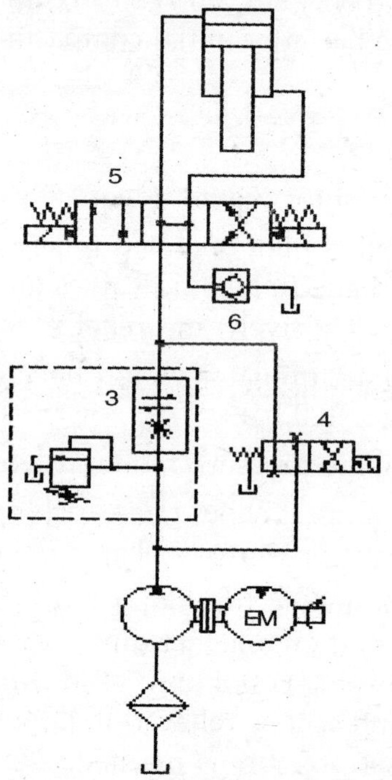

Fig. 8.11. *Rapid traverse and feed circuit.*

In the circuit shown above pump delivery normally passes through the flow control valve (3). During fast approach solenoid operated directional control valve (4) is energized. This will divert pump delivery to the cap end of the cylinder through valve (4). Full flow is thus available for the actuator to advance at the rated speed. A few millimeters before the platen makes contact with the die, the solenoid valve (4) is de-energized forcing the pump delivery to pass through the flow control valve (3). The platen now approaches the die at a controlled speed since the flow to cylinder (6) is now regulated. Directional valves (4) and (5) however must be energized simultaneously for the approach phase to begin.

Vales (4) and (5) are solenoid controlled pilot operated valves intended for handling large flows with minimum pressure drop. While valve (5) requires a 4.5 bar check valve (6) in the return line to develop the pilot pressure required to move the main spool, no such facility is required in case of valve (4) since the back pressure generated by valve (3) would serve as the pilot pressure for this valve.

8.7 RAPID TRAVERSE AND FEED, ALTERNATE CIRCUIT

In this circuit full flow from the pump is allowed to the cap end through the directional valve (3) for fast approach and the rod end oil freely passes through the normally open deceleration valve, back to tank. Near about the end of the stroke a cam depresses a roller attached to the deceleration valve spool and therefore the valve shifts blocking the flow from the rod end. The flow now has only one pathway back to tank and that is through the flow control valve (4). The approach speed is now governed by the setting of this valve. During piston retraction-stroke, full flow is allowed to the rod end through the check valve (6). Refer circuit shown in Fig 8.12.

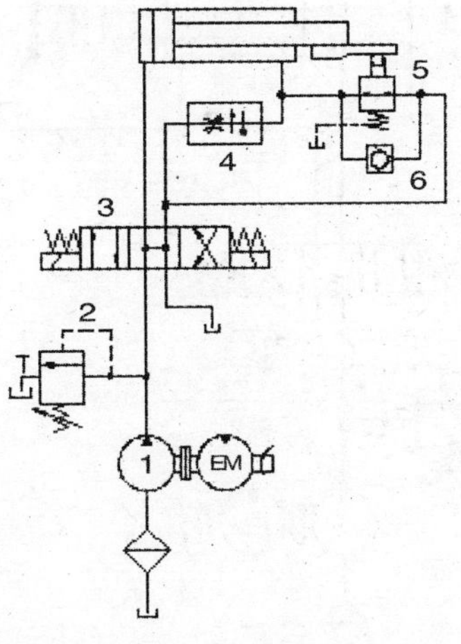

Fig. 8.12. *Rapid traverse and feed circuit.*

8.8 SEQUENCE VALVE CIRCUITS

There are applications where two or more cylinders extend one followed by the other in a definite sequence. This sequencing may be required during their forward stroke or during both their forward and retraction strokes. Before the main cylinder is actuated to perform work on a job it may be required that the job is held firmly in position. So a clamping cylinder is first actuated followed immediately by the main cylinder. With the use of sequence valves this operation becomes automatic.

In the first circuit shown below pump delivery is admitted to the cap end of cylinder (4) and the piston of cylinder (4) approaches the work piece for clamping operation. During this phase no oil is admitted to the cap end of work cylinder (6) and therefore the piston rod remains stationery. Immediately after the clamping operation the pressure increases in the line between the pump and the sequence valve (5) and as soon as this pressure reaches a value that is more than the setting of the sequence valve spring, the valve opens permitting oil flow to the cap end of the work cylinder thus actuating its piston rod.

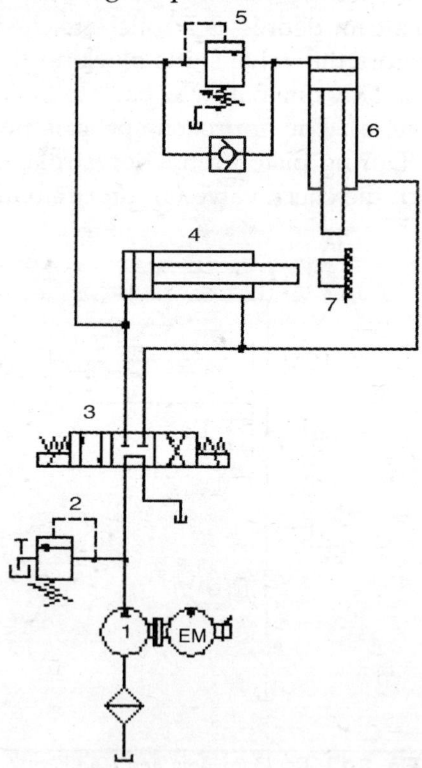

Fig. 8.13. *Circuit for sequencing of two cylinders in their forward stroke.*

It may be noted that when the work phase begins both the cylinders are subjected to the same pressure. Since the clamp cylinder usually requires a lot less force than the work cylinder the piston diameter of the clamp cylinder is proportionately less. In the second circuit (fig 8.14) the only addition is the introduction of a directional control valve (8) between the sequence valve (5) and the work cylinder (6).

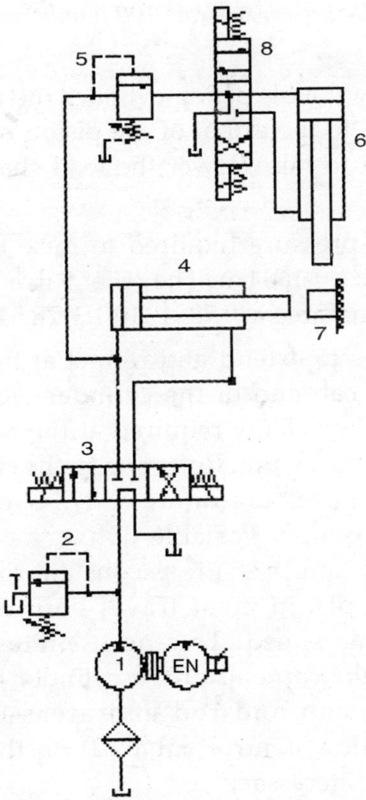

Fig. 8.14. *Circuit for sequencing of two cylinders in both forward and retraction stroke.*

The objective in this case is to achieve sequencing motion both during the approach and return strokes. The approach and return solenoids of directional control valves (3) and (8) must be simultaneously energized for sequencing of the cylinders during their forward and return strokes.

WORKED EXAMPLE

Example. *Design a suitable hydraulic circuit to raise and lower a load of magnitude 10,000 kg at a speed of 100 mm/sec. The speed must be equal both during raising and lowering of the load. The load is essentially overrunning. The load must be lowered gradually on to the platform. Calculate the flow through the* *control valves and indicate the pressure gauge readings both at the cap end and at the rod end during raising and lowering.*

Explain the reasons for your choice of the hydraulic components. Neglect mechanical and hydraulic losses. Assume 100 mm bore for the cylinder and a rod diameter = 45 mm.

Solution. Since in any double acting cylinder the rod end area is smaller than the cap end area to the extent of the piston rod cross sectional area, the pressure required to raise/lower the load shall be derived from the rod end area.

Accordingly, the pressure required to raise the load would be the 10,000/.7854 ($10^2 - 4.5^2$) = 160 bar. The relief valve setting pressure would be 175 bar. The cap end area = 0.7854 (10^2) = 78.54 cm^2.

If the cylinder has to extend and retract at the rate of 100 mm/sec flow required at the cap end of the cylinder would be (78.54 × 10) = 785.4 cm^3 /sec or 47 lpm. Flow required at the rod end would be (62.6 × 10) = 626 cm^3 /sec or 37.5 lpm. Referring to the circuit below a constant delivery pump with a pressure rating of 175 bar capable of delivering 50 lpm has been chosen. A variable delivery pump would not help because both velocity and pressure is constant throughout the cycle. It is required that the piston must travel both during extension and retraction at the same speed. This necessitates the use of the flow control valve (1) on the cap end of the cylinder because pump supply is constant but cap end and rod end areas differ. Since it is an overrunning load a flow control valve (2) on the rod end (meter-out flow control) became necessary.

Since it is required that the load must be positioned on the platform gradually, flow control (3) and solenoid operated directional control valve (4) became necessary. Towards the end of the stroke the load makes contact with a limit-switch. This energizes valve (4) to divert rod end flow through the flow control valve (3) so that the load is decelerated from100 mm per sec to 30 mm per second. This ensures the load is lowered gradually on to the platform. In order to ensure accurate speed control pressure and temperature compensated flow control valves were chosen. Flow through the flow control valve (3) during deceleration will be (62.633) =188 cm^3 /sec or 11 lpm.

While raising the load the required flow to the rod end of the cylinder is 37.5 lpm. But the pump is supplying 50 lpm. The excess flow must pass over the relief valve, which is set at 175 bar. Therefore pressure gauge P_R at the rod end will also read 175 bar. Relief valve setting pressure of 175 bar creates a retracting force on the rod end =

(175 × 62.6) = 10955 kgf. Of this 10,000 kgf is required just to balance the load.

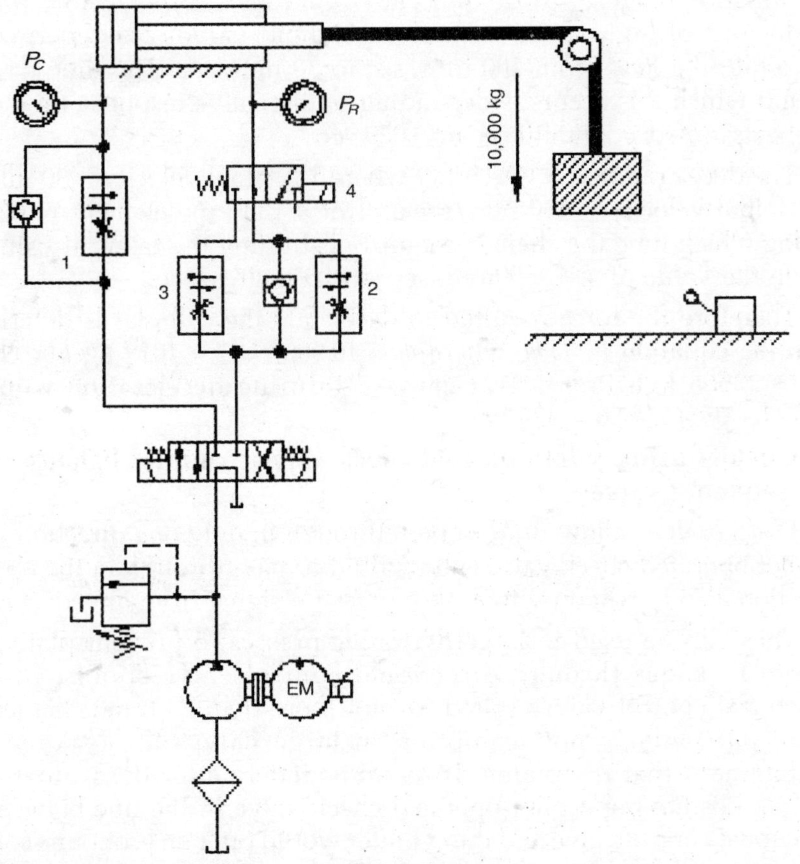

Fig. 8.15

The balance of 955 kgf acting in the retracting direction has to be balanced by the backpressure due to flow control valve (1). Consequently the pressure gauge P_c at the cap end during retraction would read 955/ 62.6 = 15 bar. When the load is lowered at a speed of 100 mm/sec the cylinder is extending at a velocity of 100 mm/sec. The flow entering the cap end of the cylinder is 47 lpm, which is less than the pump delivery, which is 50 lpm.

Pressure gauge P_C would read 175 bar because the extra flow must be dumped over the relief valve. In this operating condition the extension force (175 × 78.54) = 13,744 kgf, together with the load force = 10,000 kg is trying to extend the cylinder.

Since according to the Newton's first law of motion the net force must be equal to zero for an object moving at constant velocity, neglecting friction, the balancing force at the rod end should be 23,774 kgf. Therefore the pressure gauge P_R at the rod end would read 23,774/62.6 = 379 bar. At the end of high speed extension solenoid valve (2) is energized to decelerate the load from 100 mm/sec to 30 mm/sec. The time duration around which this occurs is depending on the valve response time which can be assumed = 20 milli-sec or 0.020 sec.

The deceleration $a = V_f - V_i / t$ where V_f = final velocity = 30 mm/sec V_i = initial velocity = 100 mm/sec and Δt = the time element = .020 sec. during which time the change occurs. Substituting the relevant values we obtain the value of a = 3500mm/sec^2 or 3.5 metres/sec^2.

Therefore the force required to decelerate the cylinder is determined from the equation $F = ma$, where m = 10,000/9.81 = 1019 kg F = (1019 × 3.5) = 23566 kgf. Pressure at gauge P_R during deceleration would be (23,774 + 3566)/62.6 = 436 bar.

Circuits using pilot operated check valves, counter balance valves and sequence valves.

Check valves allow fluid to pass through in only one direction where as pilot operated check valves allow fluid to pass through in the opposite direction also in response to a pilot pressure signal from an external line.

They can be used in large hydraulic presses to prevent platen drift due to leakages through directional control valve spools. In small directional control valves (rated for not more than 20 lpm.) this leakage is not particularly noticeable but in large valves the leakage is so predominant that the platen drifts immediately after the wall shifts to neutral. Positioning a pilot operated check valve in the line between the directional control valve and the cylinder would offer an economic solution to this problem of platen drift but as we can see there are some disadvantages. These valves however cannot prevent platen drift if the seals fail to prevent fluid leakage.

In the circuit shown in Fig. 8.16(a) it is important to determine the pilot pressure required at port x of the pilot operated check valve for it to open and facilitate return flow. This can be calculated using the following equation.

$$P_P = [(L_P - P_A)/AR_V] + [(P_A + C)/1 - (AR_C/AR_V)]$$

where, L_P is the load pressure = Load /area = 10,000/(Piston annulus area) = 10, 000 / (12.5^2 –7.5^2) × 0.7854 = 127 kgf/cm^2. P_A = residual pressure at port a of the pilot operated check valve connected to tank, considered = 0, since there is no backpressure on the check valve.

AR_v = area ratio of the pilot piston to main puppet area of the pilot operated check valve, peculiar to the check valve supplied, which in this case = 3:1. C = a constant for the check valve under consideration which can be ascertained from the manufacturer and in this case can be considered = 75.

AR_C = Cylinder cap end area / rod end area = $[12.5^2 / (12.5^2 - 7.5^2)]$ 3 0.7854 = 1.56.

Substituting these values in the equation above,

$$P_P = [(127 - 0)/3] + [(0+75)/1 - (1.56/3)]$$
$$= 198.58, \text{ say } 199 \text{ bar.}$$

We can see the disadvantage of using the pilot operated check valves to prevent platen drift in large presses since the relief valve has to be set at 199 bar instead of at 127 bar as demanded by the load pressure.

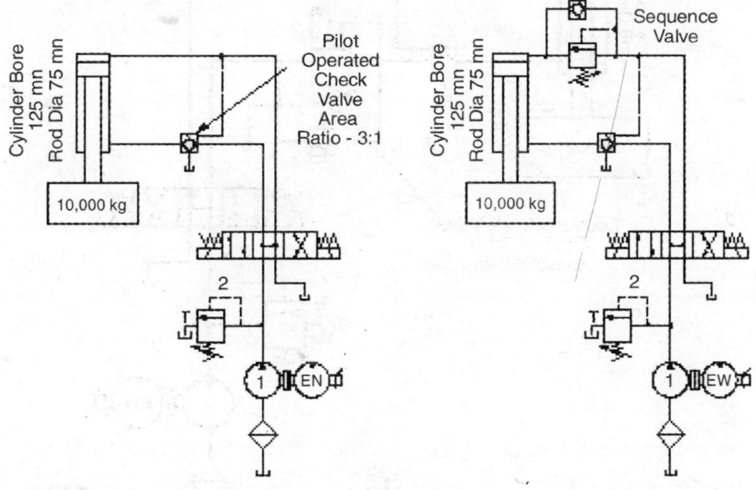

Fig. 8.16

Although setting the main system relief valve to 199 bar, would generate enough pilot pressure to open the check valve it creates excessive cap end pressure and hence rod end pressure intensification, and consequent early seal failure besides subjecting the system to de-compression shocks when the check valve opens. To overcome this problem a sequence valve in the line was included [Fig. 8.16(b)]. Since the load induced line pressure L_P in the rod end was 127 kgf/cm^2 and the area ratio of the pilot operated check valve AR_V was 3 : 1, a pilot pressure of 127/3 = 42.3 kgf/cm^2 is now sufficient to move the pilot spool of the check valve to facilitate return flow. The sequence valve was set to open at 45 bar and until the sequence valve opens there is no pressure on the cap end and at 45 bar the pilot operated check valve would have opened fully to permit free return flow.

The circuit still has a flaw in that it requires a counter balance valve to enable controlled descent and prevent the load from running away. An externally drained and externally pilot operated counter balance valve was thus included. (Fig. 8.17). The counter balance valve was set at 140 bar, which is slightly more than the load induced pressure of 127 bar.

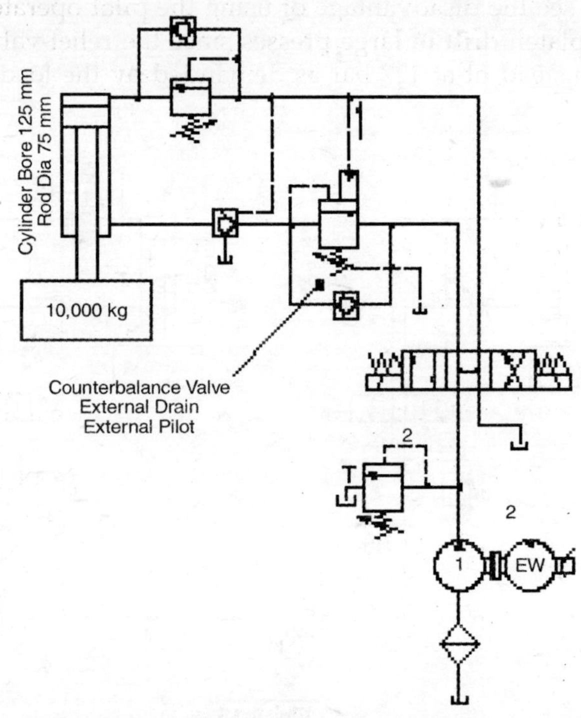

Fig. 8.17. *Pilot operated check valve circuits.*

8.9 RE-GENERATIVE CIRCUITS

In a double acting cylinder when the pump is delivering to the cap end of the cylinder oil in the rod end is expelled to tank usually through the directional control valve. If instead of diverting this oil to tank if it is made to combine with the pump delivery to the cap end, piston rod approach speed would increase to that extent thus eliminating the need for a bigger pump. But the law 'energy can neither be created nor destroyed, but can merely be converted from one form to another' is also applicable here. What is gained in 'speed' is lost in 'force'. The force developed on the cap end side is now only the balance of force between the cap end and rod end since the pressure is equal on both sides. By positioning an unloading valve on the rod end side as shown a re-generative advance with change over to conventional advance is possible thus retaining the best of both systems.

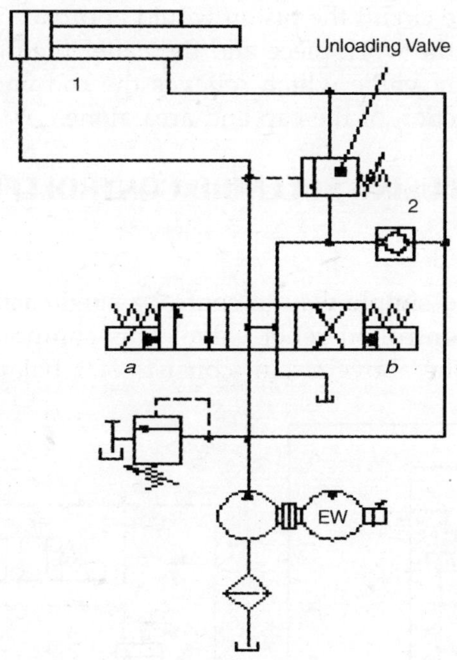

Fig. 8.18. *Re-generative circuit with change over to conventional.*

In the circuit shown above pump flow passes through the open center directional control valve to both the cap end and rod end of the actuator (1). Since the flow path to tank is also open, no pressure is generated and the actuator remains stationary. Upon energizing solenoid *a* of the directional control valve the tank line is isolated since check valve (2) prevents flow of rod end oil to tank. This flow therefore is made to combine with the pump delivery thus facilitating regenerative advance.

If the pump flow is Q flow available at the cap end during regenerative advance is $(Q \times K)$ where K is area ratio A/a. Here, A is cap end area of the cylinder and a is the rod area. The force developed at the cap end during re-regenerative advance is Ap/K, where p is the relief valve setting pressure. The re-generative approach speed is QK/A. Flow out of the rod end is $Q(K - 1)$. If we choose the rod diameter in such a way that it is one half of piston diameter, the area ratio K will become equal to 4 and the flow out of the rod end during regenerative advance will be $3Q$. To handle such large flows regenerative circuits usually have solenoid controlled pilot operated directional control valves. The cylinder and rod area ratio of 2 : 1 also makes it possible to achieve equal speeds in both the directions of piston travel eliminating the need for a double rod cylinder.

In a regenerative circuit the piston would approach at the rate of QK/A till it encounters the work piece and thereafter the increasing pressure opens the unloading valve which relieves the rod end oil to tank and thrust is now a function of the cap end area alone.

8.10 CIRCUITS USING SOLENOID CONTROLLED RELIEF VALVES

A very effective and simple way to control a single acting cylinder is by using a solenoid controlled relief valve. This component identified as component (2) in the above circuit consists of a balanced piston relief

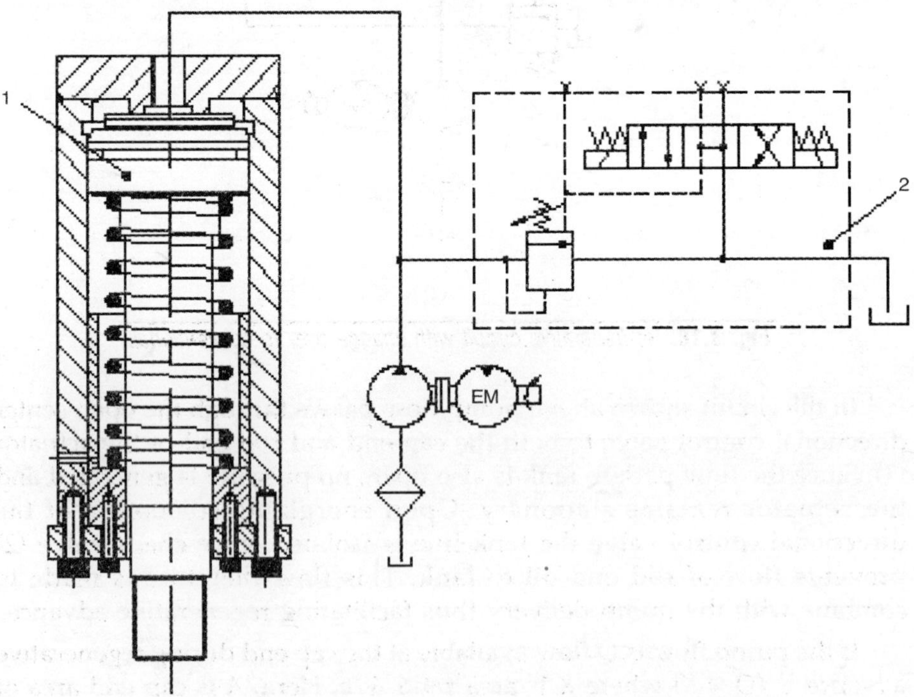

Fig. 8.19. *Circuit using solenoid controlled relief valves.*

valve on which a solenoid valve is mounted. It is available as a ready to use integral unit. Although pump delivery is directed to the cap end of the cylinder the actuator does not move until either one of the solenoids is energized.

This is because a vent line from the relief valve connects to the solenoid operated directional control valve, which in the neutral position has the vent port connected to tank. Ports *A* and *B* are plugged. Any one of these ports *A* and *B* could be used to connect a remote relief valve so that the system could respond to two different pressure signals. One as determined by the main relief valve and the other as determined by the remote relief valve depending upon, which solenoid, the right or left is engaged. This system can handle much larger flows with little or no pressure drop than the conventional systems where all of the pump delivery passes through the directional control valve.

8.11 CONTINUOUS RECIPROCATING CIRCUITS

In the circuit for continuous reciprocation of the actuator using electrical signals shown in Fig 8.20 a cam actuated limit switch can alternately energize and de-energize the solenoid of the two position 4 way directional control valve (1). In the de-energized position the actuator would move to the right and in the energized position the actuator would move to the left. The same feature can be achieved by using two 2 position/4way cam actuated directional control valves (3) and a pilot operated directional control valve (1). The pump delivery is ordinarily directed to one end of the cylinder and therefore the actuator moves to the right or left as the case may be as soon as the pump is started. The cam attached to the actuator depresses valve (3) at the end of the stroke. This diverts a part of the pump supply to the pilot pressure section of valve (1) through the flow control valve (2). This pilot pressure shifts the spool of valve (1) to permit pump flow to the other end of the cylinder, which reverses the direction of motion of the actuator.

This will be a continuous process so long as the pump delivery is maintained. The flow control valves (2) ensures that the spool of valve (1) shifts at a controlled speed by regulating the quantity of oil passing through it. Valve (1) is a pilot operated valve that is neither spring returned, spring-centered nor detented type. The spool shifts to one or the other side depending upon the pilot pressure applied and will remain in the position last attained even if the pilot pressure is removed. This circuit enables continuous and automatic reciprocation of the table without the need for electrical controls except for electric motor on/off.

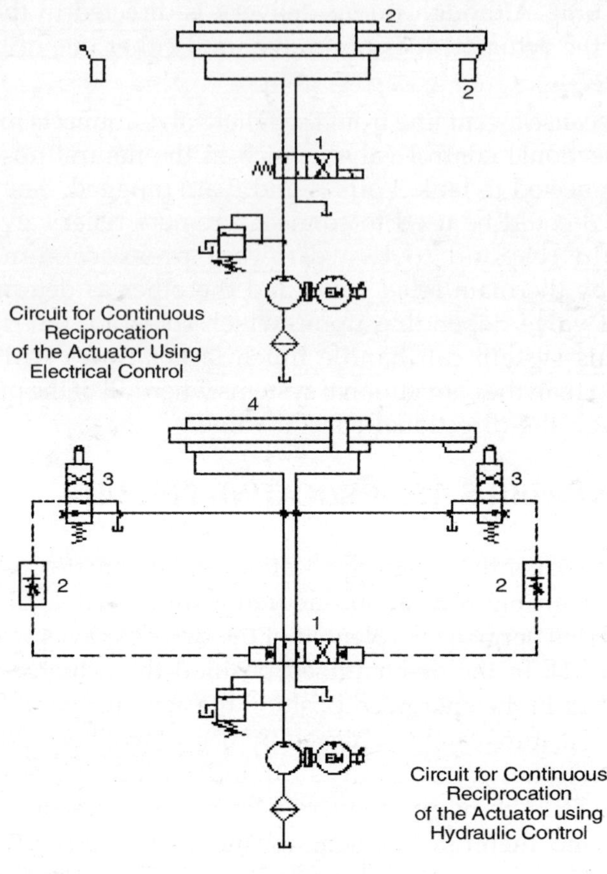

Fig. 8.20

8.12 COMBINATION PUMP CIRCUITS

8.12.1 For Variable Pressure and Variable Speeds

In the circuit shown below, normally, the combined deliveries from both the pumps is dumped to tank at low pressure because vent connections of both the relief valves (1) and (2) are open to tank through valves (3) and (4). If we energize valve (3) the speed of the actuator is controlled by pump delivery (*P*1) since (*P*2) still remains vented to tank. The higher pressure in the line from *P*1 will keep the check valve in the line *P*2 closed. If valve (4) alone is energized the actuator speed is controlled by the pump delivery (*P*2) for the same reasons.

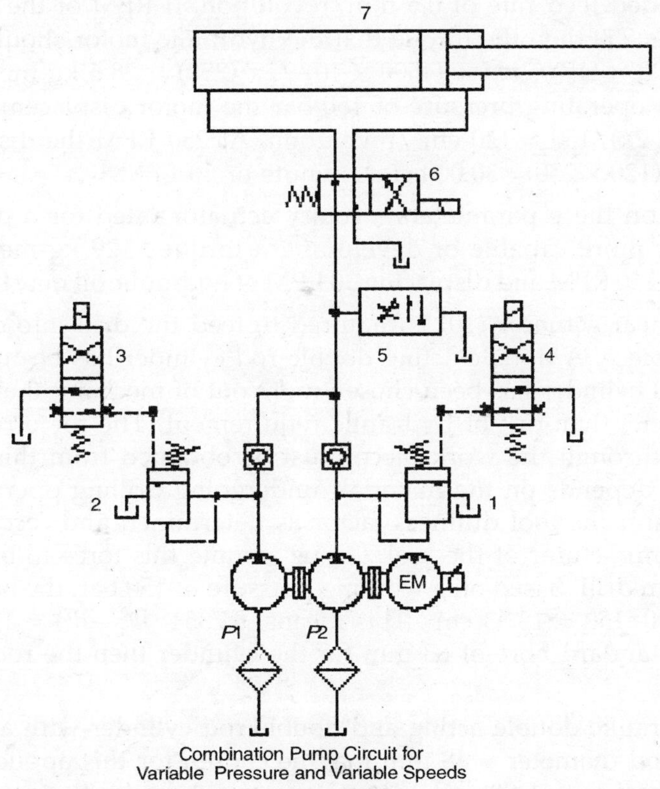

Combination Pump Circuit for Variable Pressure and Variable Speeds

Fig. 8.21

When both the valves (3) and (4) are simultaneously engaged the actuator speed is controlled by deliveries from both *P*1 and *P*2. In the above circuit if the relief valves (2) and (1) are set at two different pressures two different output forces are possible depending upon whether valve (3) or (4) is engaged. When both the valves (3) and (4) are simultaneously engaged the output force depends upon the pressure settings of valves (2) or (1), whichever is higher. The flow control valve (5) further enables infinite variation of pump delivery from zero to maximum thus making a wide range of speed control possible.

8.12.2 Circuit for Production Drilling Machine

Based on the largest drill size, cutting speed and feed rate the horsepower required for the drilling operation must first be established. This information will help size the hydraulic motor (11) that drives the drill bit. Let us suppose 10 HP is required for a 30 mm drill bit based on a recommended feed rate of 0.5 mm/revolution. If RPM of the drill bit is assumed as 250 then the torque that the hydraulic motor should develop is T = (4500 × HP)/2πN = (4500 × 10) /2π (250) = 28.6 kg-metres. If we assume an operating pressure of 150 bar the motor displacement would be (2860 × 2π)/150 = 120 cm^3/revolution. At 250 RPM the displacement would be (120 × 250)= 30,000 cm^3/minute or 30 LPM.

Based on these parameters a rotary actuator rated for a pressure of 150 bar or more capable of developing a torque = 29 kg-metres while rotating at 250 RPM and displacing 30 LPM of hydraulic oil may be selected.

The linear actuator (10) is required to feed the drill into and out of the work piece. A double acting double rod cylinder has been chosen. A double rod cylinder has been chosen more out of mechanical engineering requirements than out of hydraulic requirement. The force required to feed drill through the work piece must be obtained from this cylinder. This force depends on the material undergoing drilling operation, drill diameter and the tool dullness factor as determined and recommended by the manufacturer of the tool. Let us assume this force to be 2000 kgf for a 30 mm drill. Based on a working pressure of 150 bar, the net cylinder area = 2000/150 = 13.33 cm^2. This means, 0.7854 (D^2 -- d^2) = 13.33. If we choose a standard bore of 63 mm for the cylinder then the rod diameter $d \cong 48$ mm.

A hydraulic double acting and double rod cylinder with a bore = 63 mm and rod diameter = 48 mm may be chosen for this application. The flow into and out of the cylinder (Q) is equal because the annular areas on both sides of the piston are equal.

The recommended cutting speed for steel is 1800 cm/minute. The flow required by the linear actuator would be (1800 × 13.33) = 23994 cm^3/mt or say 24 LPM. The total flow that the pump must deliver will be (30 + 24) = 54 LPM. So a pump with a discharge rating of 54 LPM and a pressure rating of 150 bar will adequately serve both the actuator. The input HP of the electric motor is (54 × 150)/600 × 0.746 × 0.8 = 22. Two hydraulic motors (5) and (4) coupled together act as flow divider valves and ensure that each actuator gets its quota of flow and flow control valves (9) and (12) enable control of cutting speed and feed when using different drill sizes.

8.12.3 Circuit for Uniform Piston Traverse Through Varying Load

In hydraulic systems as applied to machine tools certain factors become critical which are normally ignored in the industrial hydraulic systems. For instance, in industrial hydraulic systems such as in a hydraulic press we are not particularly concerned how the piston velocity is changing with respect to load resistance throughout its travel so long as it is substantially constant. Slight variations in piston speed due to changes in the resistance to its movement and consequent higher pressure that the pump is called upon to generate and changes due to variations in viscosity are usually neglected. In machine tool hydraulics on the other hand, this variation in the speed of traverse due to higher load resistances encountered and effect of viscosity changes becomes critical.

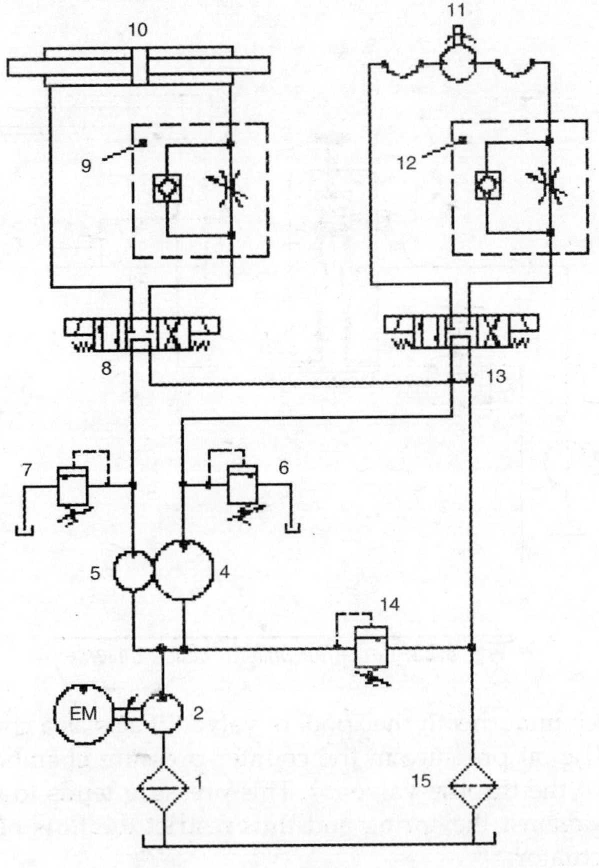

Fig. 8.22. *Circuit for production drill press.*

Variation in the speed of traverse of a milling machine table or a surface grinder would have undesirable effects on the kind of 'finish' that is imparted to the job. Various methods have been developed to ensure that for any given setting of the pump the speed is maintained constant with in close limits. One such method is discussed below.

Referring Fig. 8.23, pump (1) delivers fluid to the valve (2) via the port *P*1. The spool of valve (2) is retained in its seat by the force of a compression spring. From the primary pressure chamber of valve (2) oil flows through the port *P*2 to the directional control valve (3) from where it is directed to the cap end of the cylinder (4) which is attached to the milling machine table. Oil from the rod end is diverted to tank through the other port in the directional control valve and through port $P_3 - P_4$ and throttle valve (5) to tank. A bypass line from *P*4 is connected to tank through a by pass valve (6).

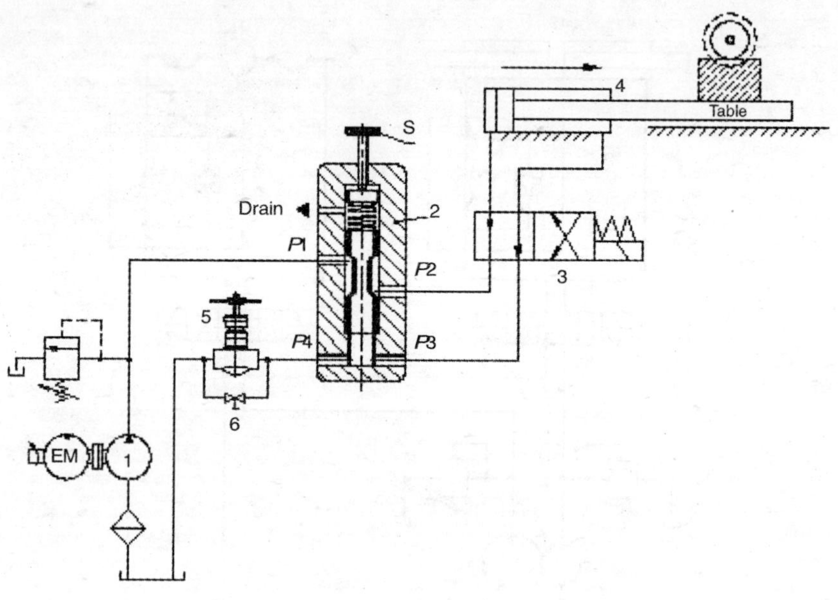

Fig. 8.23. *Circuit for uniform piston traverse.*

The space underneath the spool of valve (2) acts as a counter pressure chamber. The oil pressure in the counter pressure chamber depends on the setting of the throttle valve (5). This pressure tends to shift the spool of valve (2) against the spring and thus restrict the flow of oil from port *P*2 to the actuator.

As long as the resistance to piston travel remains constant the passage area at outlet of *P*2 remains constant and oil is delivered to the cap end of the cylinder (4) at a uniform rate. Should the resistance decrease the piston speed and consequently the rate of discharge of oil from the rod end of the cylinder tends to increase. If this discharge is more than what the throttle valve is set to allow the excess flow raises the pressure in the counter pressure chamber causing the spool of valve (2) to shift thus reducing the rate of flow to the cap end of the cylinder. If the resistance to motion increases the action is reversed. The normal position of the spool of valve (2), is set by the adjustment screw (*S*). By pass valve (6) facilitates to by pass the throttle valve if the regulating action is not required.

Problem. *A hydraulic press has a conveyor system built into it. The conveyor is used to position the die under the hydraulic cylinder for the pressing operation and return the die back to the operator as soon as the pressing operation is completed. It is required that the pressing operation should begin automatically, as soon as the die is in position and only after the piston retracts fully the die is returned to its base. Design a suitable hydraulic circuit that can effectively perform the above operations.*

Solution. Figure 8.24(a) is a schematic of the hydraulic press with the conveyor system. Figure 8.24(b) is the required hydraulic circuit.

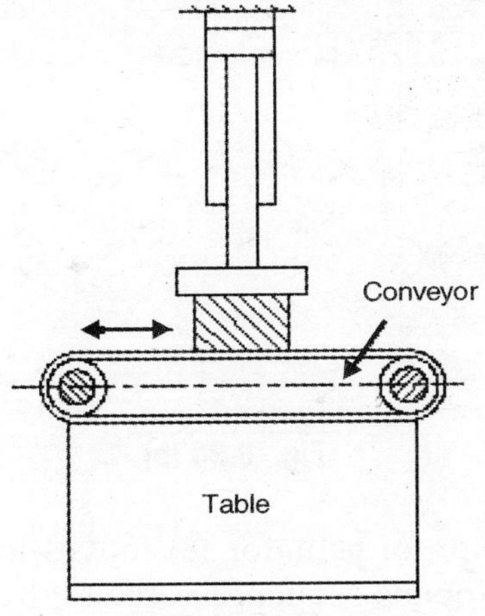

Fig. 8.24 (a)

Pump (1) delivers fluid to the solenoid operated directional control valve (2) which in its neutral position circulates oil back to the tank If we energize solenoid *a* the pump supply is diverted to the torque generator (5) and only after the plunger (7) completes its motion the oil supply is permitted to the cap end of the hydraulic actuator through the port *P*1. The plunger has racks formed on it so that the pinion, which is in contact with it can execute a rotary motion. The pinion in turn drives the conveyor. The 'rack and pinion system is so designed as to ensure the plunger stroke is just enough to position the die underneath the actuator. Check valve (3) will ensure that return oil from port *P*2 of the torque generator is directed to tank along with the rod end oil. During the return phase energizing solenoid *b* will divert pump flow to the rod end of the cylinder (6). During this phase the cap end oil is free to return to tank via port *P*1 since the plunger of torque generator (5) is still in the last attained position.

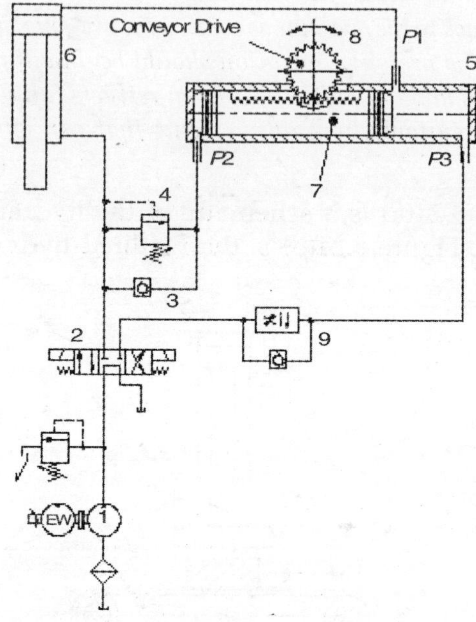

Fig. 8.24 (b)

Only after the piston of actuator (6) moves to its top most position, the pressure build-up opens sequence valve (4) to divert fluid to port (2) which moves the plunger to the right and in turn returns the die to its base. Infinite variation of conveyor speed is possible with flow control valve (9). The integral check valve permits free return flow.

8.13 PRE-FILL VALVE CIRCUITS

While discussing about various hydraulic controls, we have explained at length about the function and need for 'pre-fill valves'. Pre-fill valves are employed when the volume requirement of a hydraulic ram exceeds available pump discharge limits or where there is a clear need for reduced in-put HP and for conserving long-term energy costs without compromising on the functional requirement of the system.

Circuits employing pre-fill valves usually have one or two jacking cylinders mechanically connected to the main ram. The movement of jacking cylinders in effect causes the movement of the main ram. The function of a jacking cylinder is to carry the main ram till it encounters load resistance.

The main ram is always in direct communication with the oil reservoir with only the 'pre-fill' valve in between. Hydraulic oil assisted both by vacuum and gravity quickly occupies the space created by the movement of the main ram. This way the pump discharge required to attain a certain approach velocity need not be based on the volume requirements of the main ram, but on the volume requirement of the jacking rams, which can be as much as 25 times less! Obviously this results in enormous savings both in input HP and in long time energy costs.

If the hydraulic press is 'down acting' then it is possible to eliminate jacking rams by using a double acting cylinder. In 'down acting' presses both gravity assisted approach and gravity assisted filling of the main ram is possible.

The operation of a hydraulic press with jacking cylinder placed inside a cavity in the main ram and concentric to it is described below.

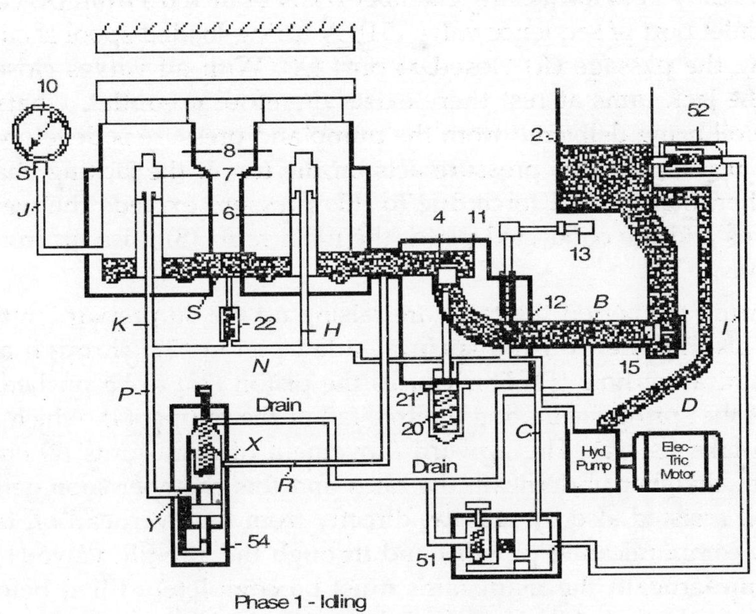

Fig. 8.25. *Hydraulic press.*

Oil reservoir (2) through the feeder line (*B*) is in direct communication with the main the main cylinder chambers (5). Control housing (11) is meant to isolate the cylinder chamber from the oil reservoir whenever required. This is accomplished through the normally open, spring actuated puppet valve (4). Control housing (11) has a screw-operated ball valve

(12) to which hand lever (13) is attached. Under idling conditions the ball valve is off its seat.

The oil reservoir has a feeder line (*D*) connected to the pump inlet line. The pump discharge line is connected to the two jacking rams (5) through control valve (11).

The discharge line from the pump is connected to the balanced piston relief valve (51) and the sequence valve (54). During idling pump delivery merely circulates through the system and returns to tank since the screw operated ball valve (12) is kept open. When the control lever (13) is turned clockwise the ball valve (12) moves down to take its seat so blocking the branch line from pipe (*G*). Now starting from pipe (*G*) at the pump outlet the oil passage path can be traced to 'pre-fill' valve (11), feed pipe (*N*) and also to pipes (*H*) and (*K*) each of these being in communication with both the jacking rams (6) which are located, in a machined bore within the main rams (8). Main rams act as the cylinder for the jacking rams. Oil from the pipes (*H*) and (*K*) enters the base of the jacking rams and issues out vertically in to the jacking chamber (7). A branch (*P*) from (*K*) connects to the inlet port of sequence valve (54). A spring-loaded spool is currently keeping the passage (*R*) closed at port (*X*). With all valves closed and with the jack rams at rest there exists therefore no-outlet, whatsoever, for the oil being delivered from the pump and pressure is developed due to this resistance. This pressure acts on the top of the jacking chambers (7). When the upward force due to this pressure exceeds the weight of the rams and the connected plates the main rams (8) raise up from their positions.

During the period the rams are raising up the oil pressure in the line (*N*) is also applied on the spring-loaded piston (21) through a small communicating hole (*F*). This causes the piston (21) to be pushed down against the spring facilitating the free fall of the puppet (4), which is now free to take its seat. The upward movement of main rams (8) creates a vacuum chamber underneath the rams and this chamber soon gets filled with oil assisted also by gravity, directly from the overhead oil tank (2) via the communicating pipe (*B*) and through the 'pre-fill' valve (11). The space underneath the main rams must be completely filled before the rams come the end of their stroke or before they encounter work resistance and before the freely falling puppet (4) takes its seat isolating the oil tank from the pressure chamber (5). Thus the chamber is maintained full of oil during jacking, but there is no pressure at this stage.

Phase 3. Pressurizing the Main Rams. When the top plate (60) during its upward motion encounters work resistance the motion stops but the pump discharge continues. The increasing pressure is now sensed at the inlet to the sequence valve (54). This pressure is communicated through an internal passage (y) to the small control piston at the base of this valve and the raising pressure causes the piston to overcome the spring at the opposite end of the spool. The latter spool is thus displaced so that its 'land' uncovers the port x so allowing the fluid to escape from this port to line (R), which is in communication with pressure chamber (5). The pump delivery is now admitted to pressure chamber (5) through the sequence valve (54). There being no out let passage for the oil being delivered the puppet valve (4) now firmly seated, pressure builds up in the pressure chamber to the setting of the relief valve (51). The pressure chamber can be de-pressurized by simply opening the ball valve (12). Because pump delivery is now allowed to the tank there being no pressure on piston (21) the spring force (20) causes the puppet (4) to open the drain passage (8) so that the rams return to their original positions due to gravity thus completing one cycle.

8.14 PRE-FILL VALVE CIRCUIT-II

This circuit is an improved version over the previous circuit. It makes use of two pumps to reduce the input power requirement and to deliver need based flow to each of the two branches of the circuit. The controls are electrical instead of manual and the production cycle is therefore automatic.

Two pumps (1) and (2) (refer prefill valve circuit–5 below) are coupled together to an electric motor. As soon as the electric motor is started delivery from pump (2) passes through solenoid valve (10) and flow control valve (7) and applies pressure on a small piston that controls the prefill valve puppet P. The oil pressure on the small piston area retracts the puppet against a spring and the oil in the overhead reservoir (13) is now in direct communication with cylinder C through the gate valve (12) and the puppet P of the prefill valve.

The oil pressure after causing the puppet P to retract blows over to tank through the relief valve (8), which is set at a very low pressure, enough to retract the puppet against a spring. During this phase delivery from pump (1) is also returning to tank through the main cylinder C and prefill valve puppet P because the passage to tank is now open.

Energizing solenoid valve (10) will divert oil flow from pump (2) to jacking cylinder J. The oil pressure acting on the piston of jacking cylinder causes it to move up also carrying the main ram R in the process. This movement stops as soon as the ram R comes in contact with the top fixed platen (F). The speed of ram movement upwards is governed by the delivery from pump (2) applied on the jacking piston area. The pressure at this point is the pressure required to overcome the weight of ram R

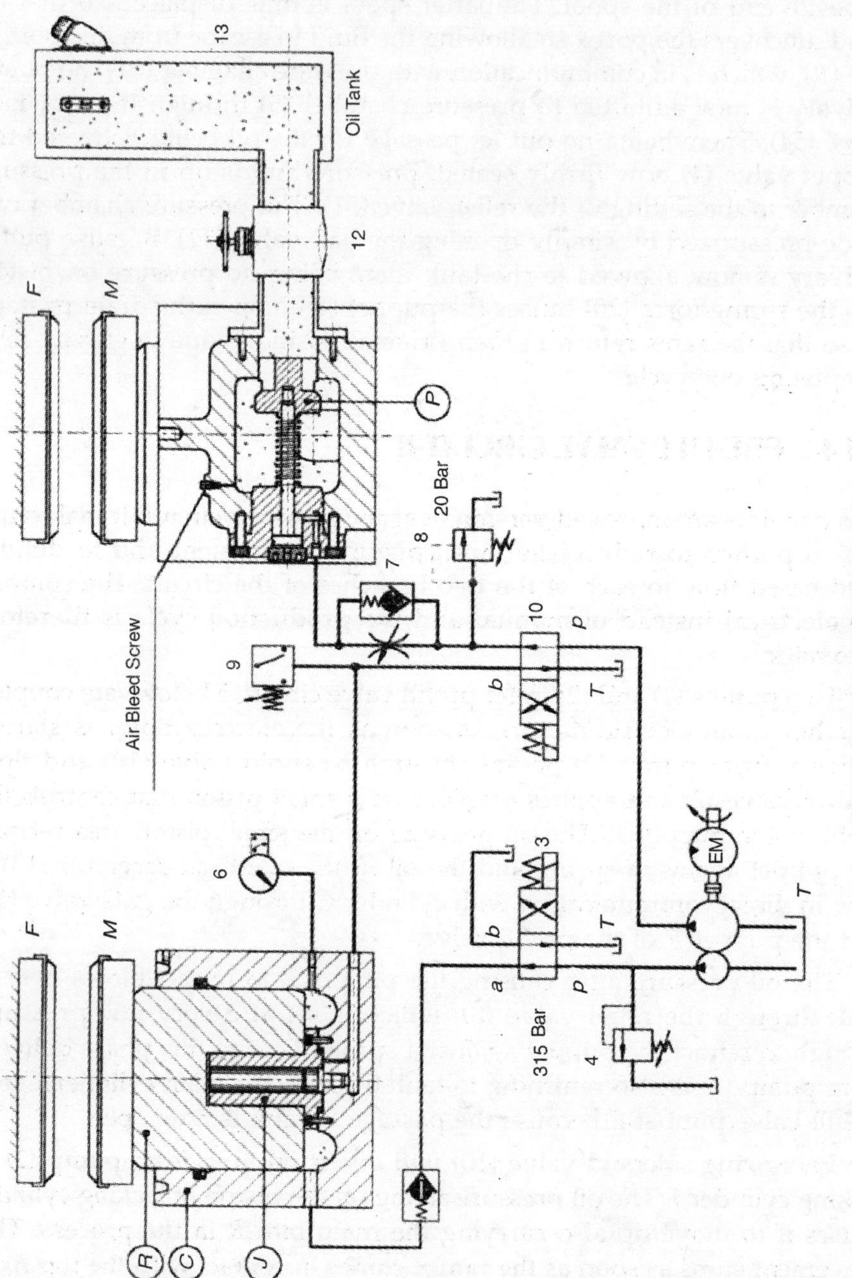

Fig. 8.26. *Pre-fill valve circuit - 5.*

and the moving plate (*M*) fixed to it. This is controlled at the pressure switch (9). As soon as the set pressure is reached, the pressure switch would de-energize solenoid valve (10) diverting pump supply. A pressure of around 50 bar on a jacking piston area of 28 cm^2 would produce a force of 1400 kg. This should be enough to jack the main ram and the plate fixed to it. With the upward movement of the main ram *R* the space created underneath is simultaneously filled with oil from the tank assisted both by gravity and vacuum. The solenoid valve (10) in its present position is directing the pump flow to jacking cylinder (*J*) through its *b* port whereas its *a* port is now open to tank. This releases pressure on the small piston that was all along holding the puppet *P* retracted against a spring. The spring now forces the puppet back to its seat to eventually seal the pressure chamber from the tank. Flow control valve (7) with integral check valve is meant to ensure restricted flow back to tank so that the puppet closing time can be controlled to synchronize with the time required for the moving platen (*M*) to contact the fixed platen (*F*).

As soon as the puppet *P* closes the tank port and isolates the pressure chamber from oil tank (13) delivery from pump (1) which hitherto was merely circulating to tank through the main cylinder *C* now has its escape path to tank blocked by the puppet *P* With no other out let available delivery from pump (1) pressurises the cylinder *C*. This pressure is indicated at the pressure gauge (6) and could be controlled at the relief valve (4).

After the moving platen (*M*) contacts the fixed platen (*F*) the consequent pressure build up at the jacking cylinder *J* is sensed by the pressure switch (9). The pressure switch is connected to an electronic timer which de-energizes valve (10) after count down. This causes pump (2) to resume its delivery and apply pressure on the prefill valve puppet piston. Since the puppet *P* is now subjected to oil pressure in the main cylinder the force acting on it is greater than the force applied through the puppet piston and hence it remains seated until the main cylinder is de-pressurised.

The depressurization of main cylinder is through the electrical contact gauge (6). As soon as the gauge senses the set pressure it signals a timer, which controls solenoid valve (3). On receiving the signal the spool of solenoid valve (3) shifts unloading the delivery from pump (1) to tank through its port *b* while port *a* and consequently the main cylinder chamber is now exposed to tank line pressure. Check valve (5) has a two mm hole drilled on its puppet to enable controlled decompression of main cylinder chamber. After the main cylinder chamber is decompressed the oil pressure from pump (2) acting on the puppet piston will retract the puppet thus permitting the ram to descend due to gravity. Thus completing one production cycle.

8.14.1 Comprehending the Use of Pressure Compensated Variable-Delivery Pumps

Variable displacement pumps, more particularly the non-pressure compensated pumps are widely used in mobile hydraulics such as in construction machinery, agricultural equipments and in material handling systems, in the closed loop circuit mode. Industrial hydraulic systems employ open loop circuits and consequently pressure compensated variable delivery pumps find their applications here.

A 200 tonne hydraulic press circuit using a pressure compensated variable- delivery pump with flat cut off compensator and with limit stop is shown in Fig. 8.27. The limit stops helps to maintain the desired flow at maximum pressure setting of the compensator spring.

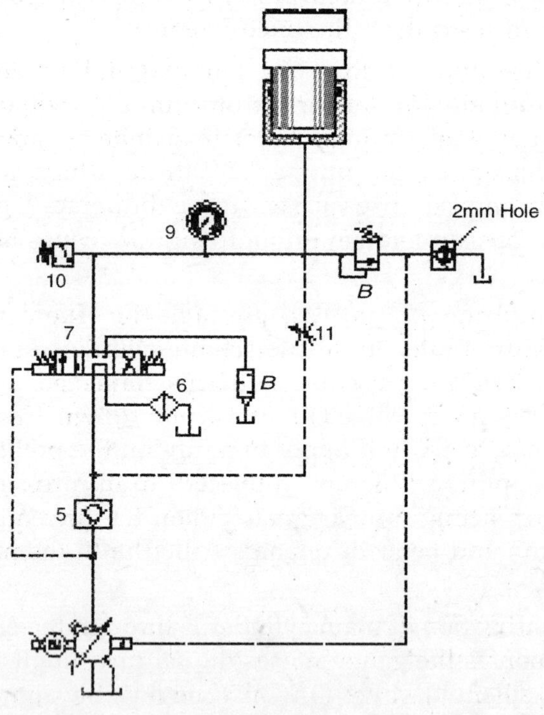

Fig. 8.27. *Circuit – 2.*

A fundamental requirement of this press was that after making contact with the die the ram should continue to move at full load but at a substantially reduced speed of one millimetre per second through a distance of 5 millimetres. This means that the pump volume control has to be applied in such a way as not to take the pump out of circuit on attainment of pressure but to leave it 'leaning' on the applied load in a live condition thus holding the pressure steady regardless of any internal leakages while consuming minimum power. A mechanical stop limits minimum pump displacement to some small value and because displacement cannot go to zero a relief valve is required to control maximum pressure.

The pump continues to deliver upto its full output untill the desired working pressure is reached. At this point the pressure-regulating valve (6) blows as a relief valve maintaining the set pressure even while bypassing the excess flow to tank through the check valve. The cracking pressure of the check valve is about 5.5 bar (85 psi), which ensures the return line is maintained at a constant backpressure of 5.5 bar. A branch from this line is given to the pump flow control plunger in the non-drive end of the pump. As soon as the oil blows over from the pressure-regulating valve the pump control plunger is pushed towards the 'no delivery' due to the 5.5 bar oil pressure that acts on the plunger.

The pump delivery is thus reduced as long as the pressure-regulating valve continues to blow at the set pressure. If the pressure tends to fall due to pump delivery going too low the regulating valve will cease to by pass excess oil to the reservoir. The 2 mm diameter hole in the check valve will drain the trapped oil in the branch line and the control plunger is forced to return to its full flow position due to the action of an internal spring. This raises the pressure until the pressure-regulating valve again by passes the excess flow. This flow while passing over the check valve to tank applies pressure on the control plunger to regulate the flow.

The dual pump circuit discussed in (Sec. 8.5 page 164) would perform the above functions but not just as well. Combined delivery from pumps (2) and (3) would maintain the required approach speed and on attaining the pressure set at the unloading valve (5) the HVLP pump would unload to tank and the delivery from the LVHP pump would ensure continued travel of the ram after attaining the set pressure. But speed at this point is based on the delivery from LVHP pump, which is fixed and so cannot adjust itself to the load requirement, as the delivery remains constant regardless of pressure. In this case it is difficult to ensure whether the ram would travel exactly at the desired rate of one millimeter per second after the load pressure is attained. By incorporating a variable delivery pump we have ensured that the flow is regulated to the desired level by adjusting the pressure relief valve (6). The pressure relief valve controls both the operating pressure and the flow to the desired value.

We have eliminated the dual pump arrangement, the unloading valve (5) the check valve (4) used in the dual pump circuit. The only additions are, a relief valve and a simple check valve. Input power requirement in the case of dual pump circuit was based on the HVLP pump delivery and maximum pressure setting. In case of variable delivery pump the input power requirement is based on the full flow rating of the pump and maximum pressure setting. Obviously the input power requirement increases many times over although energy consumption would be same in both the cases.

A better alternative to Circuit-2 is Circuit-3 shown below.

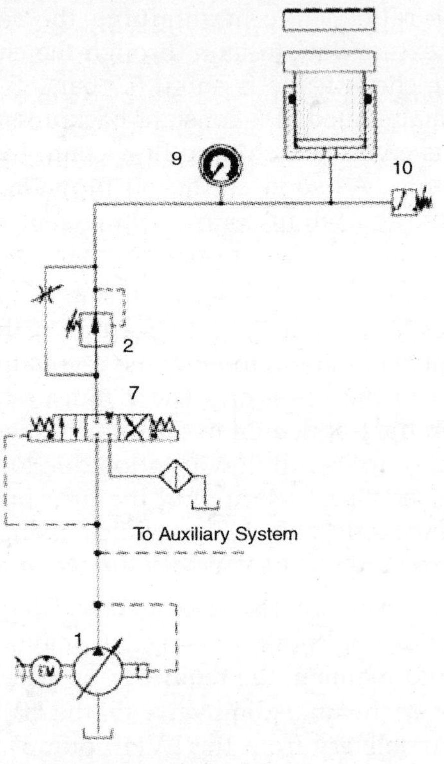

Fig. 8.28

Here the directional control valve has a closed center and the pressure signal to the compensator is taken from the main pressure line and we have eliminated the check valves and the muffler (8) altogether. But a separate de-compression valve (2) was required in view of the closed center spool configuration of the directional control valve. Pump characteristics are similar to the pump used in Circuit (2) but this pump has a flat cut off compensator without limit stop. This characteristic of the pump has further eliminated the need for a system relief valve, since in this case pump delivery can go to zero at maximum pressure setting of the compensator spool.

As soon as the electric motor is switched on delivery from the pump is blocked at Valve (7). The pump regulates its flow as soon as the compensator pressure setting is reached. The line between the pump and the directional control valve is now at the maximum system working pressure. Besides simplicity the other advantage of this circuit over the previous circuit is that an auxiliary line from the pump can be taken to perform some other function or to run a similar machine if the duty cycle permits.

In Fig. 8.29 below pump (2) is a pressure compensated variable delivery pump with a flat cut of compensator without limit stop. This means the pump delivery will be zero at maximum system pressure. Therefore the system needs no relief valve. Directional control valve (5) when energized will allow oil from the pump to either the cap end or the rod end of the cylinder (8) depending on whether the LH or the RH solenoid was engaged. The piston rod accordingly moves up or down at the speed governed by the maximum pump delivery.

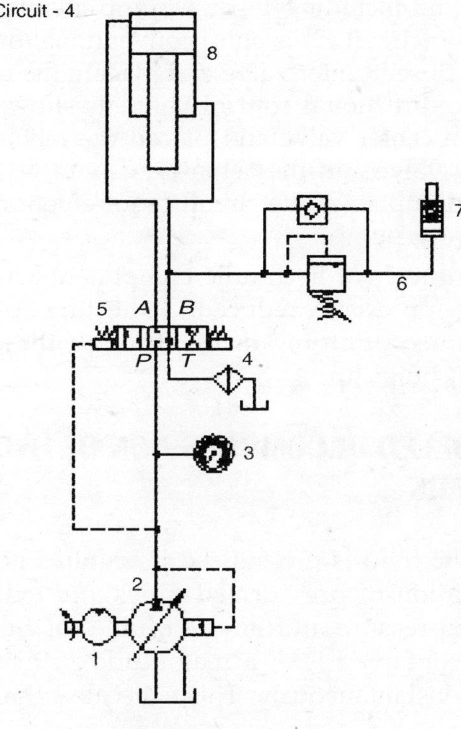

Fig. 8.29. *Circuit – 4.*

In this press it was required that the die should remain closed for 10 minutes under full pressure and no pressure variation was permitted during this period. This therefore is an ideal application that justifies the use of a variable delivery pump. The energy consumption during the 10 minutes dwell time is technically zero although the pump is holding the pressure during this period by making up for pressure loss due to leakage past the seals or the directional control valve.

After the dwell time the directional control valve is shifted which allows pump delivery to the rod end of cylinder (8). After the piston rod withdraws to its original position the pressure increase causes sequence valve (6) to open and this permits pump flow to the ejector cylinder (7) and causes ejection of the pressed component from the die.

After ejection, the direction control valve (5) is shifted to neutral position. Trapped oil in the ejector cylinder is returned to tank through check valve (9), as the line is open to tank in the valve neutral position. The spool center configuration chosen in this case is different from the previous two cases. In the neutral position *P* and *A* are blocked while *B* is open to *T*. Although *B* is open to *T* the piston would still not drift because *A* line remains blocked.

All of these circuits including the dual pump circuit explained in *demand flow-open circuits* but circuit (2) is an unconventional one in the sense that instead of using a closed center valve and placing the relief valve between the pump and the directional control valve as shown in circuit (3), we have used an open center valve and placed the relief valve between the directional control valve and the actuator. Circuit (4) justifies the use of variable delivery pump since the circuit requirement necessitates the use of a variable delivery pump.

Besides, the number of hydraulic components required and piping complications gets appreciably reduced. Oil heating up and wasted energy is also reduced since dumping of oil through the open center valve contributes to heat build up.

8.15 CONTROLLED DECOMPRESSION OF HYDRAULIC CYLINDERS

When the hydraulic fluid is pressurised it acquires potential energy due to elastic deformation of pressurized pipes and cylinders. In systems where both 'high pressure and large volume' of fluid is involved the potential energy contained is of a magnitude sufficient to damage the system if released instantaneously. This is because the rate of return flow through the control valve increases several times over the normal. A large portion of this energy is dissipated as pressure shocks and the rest as transient noise. Pressure shocks can damage pipelines and other system components and in extreme cases can endanger the personnel nearby. In the return circuit filters or oil coolers may be damaged or destroyed due over pressurisation.

In Fig. 8.27 above we have eliminated this problem by placing a needle valve (11) in the line. This will enable controlled de-compression of the oil when the directional control valve (73) is shifted to neutral position. If the valve is solenoid operated a pressure switch may be used to signal the valve to neutral after the set pressure is reached.

In the valve neutral position the pressurised oil in the cylinder is constrained to escape to tank only through the orifice thus restricting sudden release of energy. After the pressure is released the valve is shifted again to permit gravity return of the ram.

Another method of decompression of hydraulic cylinders is by using a de-compression valve (2) as shown in Fig. 8.27. This valve permits free passage of oil to the cylinder till the set pressure is reached and then closes the passage. If the directional control valve (7) is now shifted the return oil from the cylinder has only one escape path to tank and that is through the restricted passage in the decompression valve. After the cylinder is decompressed the decompression valve opens to permit free flow from the cylinder to tank enabling the ram to descend.

Un-controlled deceleration of a heavy moving mass connected to a linear or rotary actuator also results in pressure shocks. One of the methods employed to control pressure shocks is by the use of cross line relief valves placed as shown in the decompression circuit in Fig. 8.30.

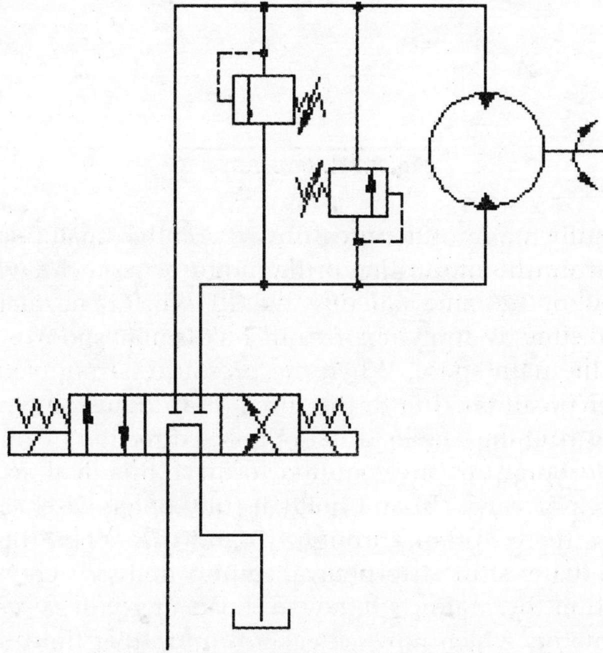

Fig. 8.30. *Decompression circuit.*

However this method is suitable only for rotary actuators and for double rod end cylinders with equal areas on both sides.

Special circuits called *brake circuits* using specially modified counter balance valves (valve *B*) are designed to stop a heavy moving mass with minimum of shock. A brake circuit is also employed when the load is 'over-running.' Figure 8.31 illustrates how the brake circuit functions. The brake valve (*B*) has a large spool and a pilot spool.

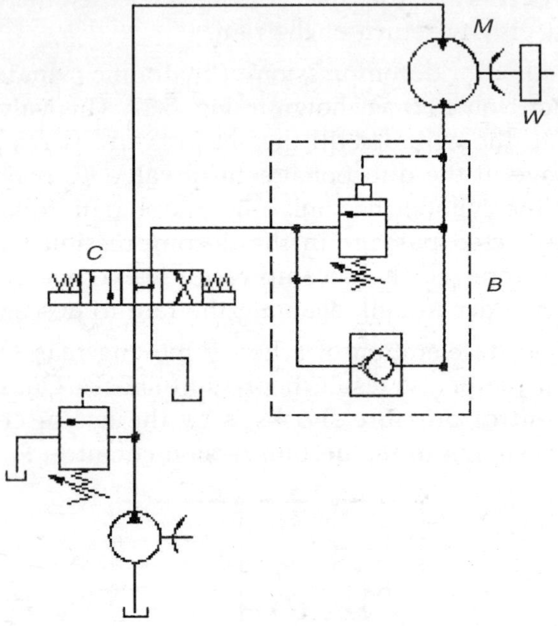

Fig. 8.31. *Break circuit.*

The hydraulic motor outlet pressure acts on the small pilot spool and the pressure from the input line of the motor acts on the whole area of the large spool on the same side of valve (*B*) which is normally closed. It can be opened either by applying pressure on the pilot spool or by applying pressure on the main spool. When the pressure is removed the spools return to their positions due to the force of an adjustable compression spring. During running, the load (*w*) opposes direction of rotation of the motor (*M*). Working pressure required to drive this load acts under the large spool area of valve (B) and holds it fully open. Discharge from the motor returns freely to tank through (*B*) and (*C*). When the directional control valve (*C*) is shifted to neutral, pump delivery gets diverted to tank and load on the motor is removed. Load inertia however continues to drive the motor, which now acts as a pump. Inlet fluid to the motor (*M*) is supplied through the valve (*C*). Since the pressure on the large spool of valve (*B*) is also withdrawn due to shifting of valve (*C*) to neutral, the large spool spring returns which restricts discharge from valve (*C*) creating a, back pressure at its outlet thus controlling the rate of deceleration.

8.16 APPLICATIONS OF ROTARY ACTUATORS

In industrial hydraulic applications hydraulic motors have effectively replaced electric motors and the mechanical clutch and gearbox assembly especially where high torque transmission and infinite speed control facility is required. Besides being compact and highly efficient they do not suffer from problems such as clutch plate wear out and gear tooth wear out. Hydraulic motors however are more widely used in mobile hydraulic systems for hydrostatic transmissions in the closed circuit mode.

Since a discussion on mobile hydraulic systems is outside the scope of this book, we will limit our discussions to the application of hydraulic motors to industrial hydraulic systems.

In Fig. 8.32 roller (11) is driven by an electric motor. Roller (12) is an idler and is free to rotate on its bearings. Roller (13) has to transmit the load torque and should be capable of rotating in either direction with infinite peripheral speed adjustment facility within a given range. This roller is carried on the support bracket (14), which is pivoted at *M* and at *N* to the connecting rod (15). The connecting rod has an eccentricity *e* about its center to which the drive shaft (16) is keyed. The rotary motion of the drive shaft (16) is converted in to reciprocating motion of the connecting rod (15) which enables bracket (14) to execute a circular motion about its pivot *M* to facilitate opening or closing of roller (13) with respect to roller (11).

Example. *Devise a suitable hydraulic system to drive roller (13) and shaft (16) that meets the above requirements. Roller (13) is required to rotate in anti-clockwise direction only after it makes contact with roller (11) and reverse its direction of rotation after the contact is withdrawn.*

Solution. Pump (2) draws its oil from the tank through inlet strainer (1) and this supply is normally delivered straight to port C of the torque generator (5) through the four way two position, spring off-set, lever operated, directional control valve (4). The oil pressure drives the piston *S* of the torque generator to one extreme position which coincides with the opening of port *d* to permit oil delivery to the *g* port of the hydraulic motor (10) through lever operated directional control valve (9). This enables the hydraulic motor to execute anticlockwise rotation at a speed controlled by the pump delivery. The expelled oil from the hydraulic

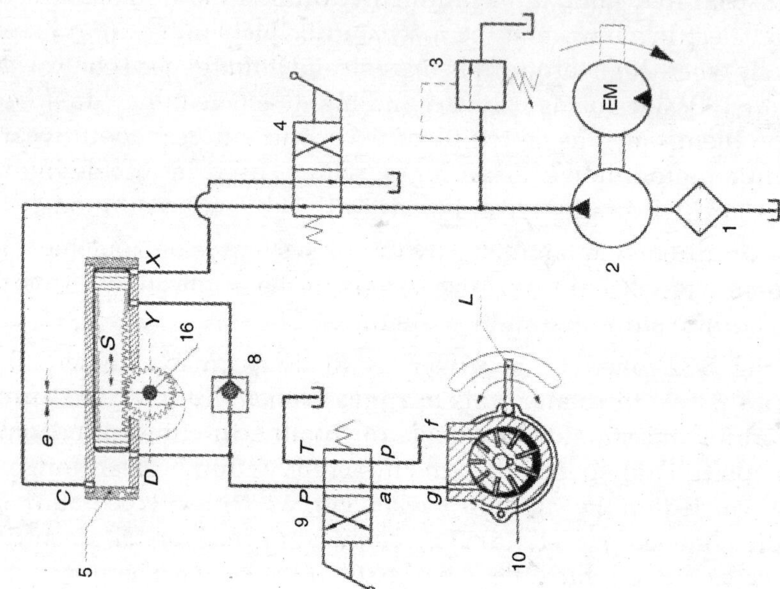

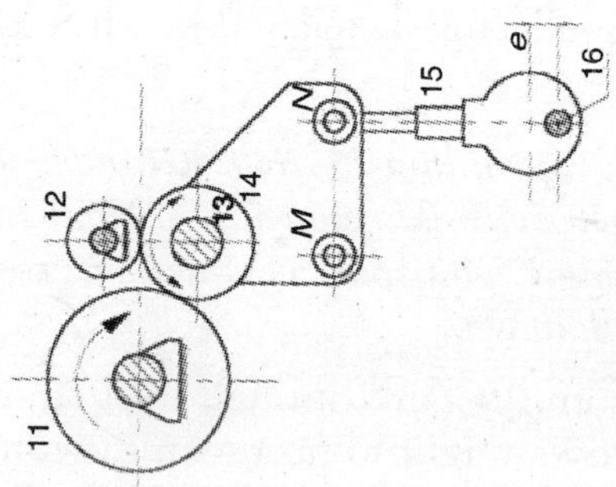

Fig. 8.32. *Hydraulic circuit.*

motor is returned to tank through the *b* port of valve (9). The hydraulic motor is meant to drive roller (13).

The hydraulic motor is of the unbalanced vane type with variable eccentricity for regulating the rotational speed. The speed range of the motor has a ratio 1:1.5, which means the maximum speed is 50% more than the minimum speed. Altering the position of lever *L* can vary the rotational speed of the motor.

The piston *S* of the torque generator (5) has gear teeth cut on it and is mesh with the pinion (6). The 'to and fro' motion of piston *S* causes a rotary motion of pinion (6). The pinion is intended to drive shaft (16). If we shift hand lever of valve (4) the pump delivery is now admitted to port *x* of the torque generator (5).

This causes piston *S* to move to the other extreme causing withdrawal of roller (13) away from roller (11). After piston *S* attains its final position it uncovers port *y* which permits delivery to the hydraulic motor through check valve (8). Actuating valve (9) will cause clockwise rotation of the vane motor, thus fulfilling all the requirements.

8.17 APPLICATIONS OF INTENSIFIERS

Problem. *A certain hydraulic press is designed to operate at a maximum pressure of 100 bar. It is required to modify the hydraulic circuit of this press so that it is also capable of operating at 200 bar whenever required. Since this high pressure is required only at random intervals cost considerations must be borne in mind while designing the circuit. It is assumed the machine is structurally sound and can resist higher stresses. The 200 bar pressure is required on a cylinder of bore 200 mm over a stroke of 25 mm.*

Solution. The high-pressure requirement could be met by any of the following three additions or alterations to the circuit.

- By replacing the existing pump with a new pump, which can build up pressures up to 200 bar. But since the input HP requirement also increases correspondingly, a new electric motor would be an additional requirement.
- Replacing the existing cylinder with a new cylinder having a bigger bore would solve the problem but since the pump delivery remains the same it would affect the speed of the actuator and therefore the productivity of the machine.
- The other alternative is the use of an intensifier. Since the pressure intensification required is equal to 2:1 the intensifier would be cost effective. Also it is simple to manufacture and can be assembled 'in-house.' In some cases a discarded or out of service hydraulic

cylinder could be converted into an intensifier with minor modifications.

In this case since the bore is 200 mm and since we need high pressure for just about 25 mm of the stroke, we need $(20^2 \times 0.7854) \times 2.5 = 785$ cubic centimeters of hydraulic oil at 200 bar.

If we use an intensifier of bore 150 mm and rod diameter =110 mm it would give a pressure intensification of 2.16:1 which is acceptable. If such an intensifier has a stroke of 90 mm it would displace $(15^2 - 11^2) \times 0.7854 \times 9 = 812$ cc of oil at 200 bar and this meets the requirement.

The modified hydraulic circuit with the intensifier is shown in Fig. 8.33.

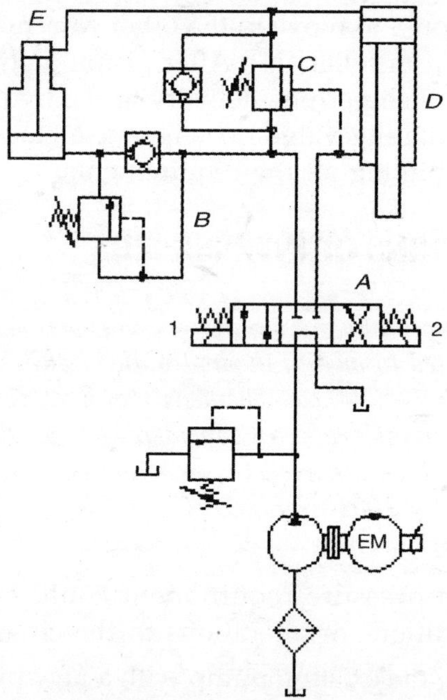

Fig. 8.33. *Intensifier circuit.*

The operation is as follows.

The operator energizes solenoid (1) of directional control valve *A*. This directs the oil flow to the cap end of the press cylinder (*D*) via the by pass check of sequence valve *C*. Oil is also directed to the rod end of the intensifier (*E*) and to the primary port of sequence valve *B*. The sequence valve *B* is set to open at 110 bar. So long as the system requirement is within this limit sequence valve *B* remains closed permitting normal operation. When more than 110 bar is required the sequence valve *B* opens permitting oil flow to the cap end of the intensifier cylinder (*E*). High-pressure oil from the rod end of intensifier (*E*) now flows to the cap end of press cylinder (*D*) to perform the required task at 200 bar. The press cylinder (*D*) can be retracted, by energizing solenoid (2) of valve *A*. Sequence valve *C* will ensure that the returning oil from the cap end of the press cylinder *D* will first retract the extended intensifier before the sequence valve *C* opens to permit return oil to tank.

9

MAINTENANCE OR TROUBLESHOOTING

9.1 INTRODUCTION

The commonly used phrase 'maintenance and troubleshooting' did not make such a good sense to me and so I have worded it slightly differently only to emphasize the fact that it is poor maintenance that often leads to frequent troubleshooting. Proper and scheduled maintenance should eliminate the need for trouble-shooting. This topic on the maintenance of hydraulic machinery pre-supposes the fact that the concerned machine is basically well designed and reliable and that the concerned maintenance personnel are well trained and motivated. The crux of the problem of course is the availability of maintenance personnel especially those who are well trained and well exposed to hydraulic machinery maintenance.

It is not the purpose here to lecture on the already well-documented theories regarding the various techniques adopted for maintenance of machines in general. It is assumed the reader is well aware of it. Here the discussion is confined to the problems, peculiar to hydraulic machines which if not attended to, will cause gradual deterioration in performance or sudden machine breakdown.

One of the vital documents that the concerned maintenance personnel should be familiar with is the hydraulic circuit diagram. He must able to read the diagram and must be able to comprehend the function of each part before attempting to solve any problem as otherwise the problem may get compounded. Another important document is the log sheet that should be maintained for each machine wherein all pertinent information regarding the machine are recorded and continuously updated. Vital information like, the name of the machine, the name and address of the manufacturer, the year of purchase, complete specification of critical spares, tonnage of the machine, the maximum relief valve setting pressure, the cycle time of the machine, oil tank capacity, quantity and specification of oil, especially the oil viscosity and oil temperature after 8 hours of running etc., must be recorded in the log sheet.

Details regarding how and when a certain problem manifested itself and what remedial measures were undertaken to correct the problem must be recorded. All this, would help analyze any deterioration in performance that might have occurred over the years and take corrective measures.

The biggest problem that one experiences while engaged in the troubleshooting of hydraulic machinery is the non-availability of diagnostic tools at the spot and the absence of pressure and flow tapping points at critical locations. The basic diagnostic tool is the pressure gauge and the one fixed on the machine will invariably be out of action in 50 per cent of the cases.

Even big companies do not carry a simple industrial thermometer that is required to measure oil temperatures, in their stock, not to talk of flow-meters and tachometers. A tongue tester, which measures the current drawn by the electric motor can act as an important diagnostic tool as it can help evaluate the condition of the pump and the condition of the electric motor itself based on the current that the electric motor is drawing. If the electric motor is getting heated up to more than normal values we do not know whether it is due to a bad pump or due to an inefficient electric motor or due any other problem and so it becomes important to eliminate the pump and the electric motor.

Provision for mounting a pressure gauge and a flow-meter at the discharge end of the pump upstream of the directional control valve will help evaluate the condition of the pump. Other important parameters that requires, monitoring, are acceleration and torque. These parameters can help evaluate the condition of rotary actuators. In the absence of these diagnostic tools it is mostly the experience, the sense of touch and the sense of hearing combined with a little of wisdom that usually enable a service engineer to diagnose and troubleshoot. This is not to decry however, the importance of these faculties of mind.

Well designed and proven hydraulic machines unlike mechanical machines require little or no maintenance for a long time if at the time of installation and commissioning, manufacturers guidelines were strictly followed. In hydraulic machines there are no lubricating points and so no daily maintenance is required. Any problem that may crop up may mostly be due to abuse, negligence, voltage fluctuations, inadequate air circulation around the machine, very high or very low ambient temperatures, improper adjustment of control valves etc., all attributable to extraneous factors.

Let me quote a few instances, in support of the above statement.

A certain customer once called me and requested that I should help him solve a certain problem with his hydraulic machine. When I did get there, I saw that they had dismantled the machine completely in their eagerness to solve the problem. I told the customer that in this condition I will not be able to diagnose the problem and so requested him to re-assemble all the parts. After assembly I asked the maintenance engineer to start the machine. The moment he started the machine I diagnosed the problem. The electric motor was running counter clockwise considered from the shaft end of the pump. It appears a day earlier the maintenance personnel while attending to an electrical problem in some other machine next to it inadvertently changed the electrical phase for this machine with the result the motor was running counter clockwise, while the pump was assembled for clockwise rotation.

Gear pumps deliver fluid only in one direction that is, clockwise when observed from the shaft end. Vane pumps can be assembled at the convenience of the customer for delivery of fluid in either the clockwise or counter clockwise direction. Piston pumps deliver fluid in either direction.

I got a similar call from a different customer for more or less a similar problem. The complaint was the piston is not moving up. Clogged inlet strainer was the cause. The pump was not drawing any supply from the tank and so it cannot be expected to deliver any fluid. After thoroughly cleaning the strainer in kerosene and by compressed air the machine was back to normal.

Another complaint that I received was in respect of a machine, which was working without a complaint for two years but suddenly developed problems. The electric motor was stalling and was getting heated up too much within half an hour of operation. It was a dual pump system with an unloading valve, which was supposed to relieve the high volume pump at 30 bar. But somebody had altered the setting of this valve and so the un-loading valve was being relieved at 100 bar instead of, at 30 bar. So the motor was getting over loaded, was heating up and therefore stalling.

Another customer had problems with his machine because he was running it for 24 hours in three shifts. The machine next to it was a plastic extruder with high wattage heating system and this was keeping the room temperature much above the ambient. Oil, in any hydraulic system heats up when work is done on it. Most of this heat dissipates to the surroundings through convection and the rest is what causes the raise in oil temperature. So long as this temperature stabilises at some acceptable value there is no problem. But if the room temperature itself is above normal, it affects the rate of heat dissipation and this in turn affects machine performance. The customer failed to appreciate this but argued that another machine next to the hydraulic machine was working fine under the same environmental conditions and for as many number of hours. I explained that it was a pneumatically operated machine where the air after doing work is expelled to the atmosphere and fresh air is drawn in at the start of every cycle whereas in a hydraulic machine the same oil circulates.

One hour of rest for every eight hours of operation is recommended for all hydraulic machines working without heat exchangers especially if the ambient/room temperature is above 30 degrees centigrade.

I have narrated the above incidence only to highlight the fact that more often than not, it is the human factor that is the cause of the hydraulic system failure and the clue to solving the problem lies in the fact that all of them were working fine for years without a problem.

Troubleshooting in hydraulic machinery, is similar to identifying the 'cat among the pigeons'. Among the various hydraulic components in the circuit usually one would have failed and the task involves identifying this component starting from the inlet strainer to the actuator, by a process of elimination.

Noise and heat are the early warning signals that the hydraulic service engineer should take note of. Inadequate fluid in the reservoir, viscosity of the fluid being too low or too high, air leak in the suction line, improper alignment of drive shafts, clogged filter elements, improper adjustment of relief valves are some of the problems associated with heat and noise in any hydraulic system. If the electric motor gets heated up to more than permissible temperatures within a short time it is usually an indication that the pressure control valves are jammed or, are set to too high values. A worn out pump also generates noise and heat due to low volumetric efficiencies.

Two of the most frequent complaints in any hydraulic machinery are, it takes 'too long for the set pressure to develop and the actuator speed is slow. Both of these problems if present at the same time, it can be attributed one single source, a worn out or damaged hydraulic pump with reduced volumetric efficiency due internal leakage. Such a pump will get heated up within half an hour of operation and will also be making more noise than normal. The electric motor coupled to the pump may be overloaded and therefore will be working at higher temperatures than normal. Pump delivery alone is no indication that the pump is in good shape. The pump must maintain the rated delivery at the rated pressure within the limits of its volumetric and mechanical efficiencies. Working on a continuous duty of 8 hours a day at the rated pressures, gear pumps wear out in about 2–3 years, vane pumps in about 4–5 years and piston pumps in about 5–7 years. The life spans indicated is based on the assumption that the hydraulic oil used was of proper specification, was free of dust and ferrous particles and the pump and the motor shaft were in perfect alignment. If it can be assessed that the pumps have not been worked for that many number of hours as yet and other symptoms associated with pump failure are not present, then it can be concluded that the pumps are in good shape and the source of the problem is else where. In a double acting cylinder if the piston seals are damaged then the leakage is internal and hence difficult to detect. A damaged piston seal in a double acting cylinder would allow fluid transfer from cap end to rod end resulting in increased pressurization time and slow piston movement. If on the other hand rod seals are damaged the leakage would be visible externally and corrective action can be taken immediately.

While troubleshooting, it is important to distinguish the problem clearly. For instance slow actuator movement and reduced pressurisation time means that there is flow but at reduced rate and the pump can build up pressure up to the limit of the relief valve but takes a long time to reach the set value.

This clearly shows that the fluid is draining somewhere either within the pump itself if the pump seals are damaged or within the actuator. If for instance the relief valve was set to a lower value the pressurisation time will not be affected but the maximum pressure developed would be limited to the setting of the relief valve. Similarly if the flow control valve is set to pass very little fluid then the speed of the actuator would be affected but it neither affects the pressurisation time nor the maximum system pressure.

Priming of pump means displacing air in the pumping chamber completely and replacing it with hydraulic oil. In case of pumps mounted below the oil reservoir this can be accomplished easily by loosening the pump discharge line and allowing the entrapped air to escape. Soon oil starts oozing out of the discharge port indicating that all the air entrapped is displaced. In case of pumps mounted above the oil reservoir the pumps must be driven in the right direction first before it is ensured all the air is driven out of the discharge line and the pump is primed. Priming of pump every time oil is replaced especially, for swash plate piston pumps, which have a large pumping chamber, is recommended to prevent entrapment of air in oil and subsequent problems.

If the pump is properly primed, if there is enough oil in the reservoir and the pump is turning in the right direction you can be certain that the pump is delivering the fluid because even a worn out pump delivers fluid. If this delivery is producing no results on the actuator then all of this delivery must be by passing to tank through the system relief valve which is struck open or set to too low a pressure or the pump is damaged and therefore incapable of building up 'pressure' due to internal leakage.

Solenoid valves are very sensitive to dust and dirt and voltage fluctuations. Minute foreign particles getting lodged in the valve spool clearances could result in the failure of these valves. All solenoid valves have a manual over ride provision and by manually pushing the spool with an appropriate tool it can be ascertained whether it is the spool problem or the solenoid problem, the only two sources for the solenoid-operated valves to malfunction assuming of course the compression spring which returns the spool to its original position is intact. If the problem is due to solenoid 'burn out' then the spool would have returned to its de-energized position but if it is due to dirt, the spool will 'stick' in the last attained position. If it is a case of solenoid 'burn out' the valve and therefore the actuator would not respond to electrical signals. In such cases the actuators would respond to manual displacement of solenoid spool for which a provision exists in all solenoid valves. Even if the spool is struck due to foreign particle embedded in the valve spool clearance the valve would not respond to electrical signals.

In such cases complete disassembly of the valve and cleaning of valve spool and housing in kerosene and compressed air would solve the problem. Before putting the valves back in position ensure that the holes in the valve body and those in the sub-plate matches because some valves are not provided with dowel pin location.

Hand operated directional control valves rarely present problems. Solenoid controlled pilot operated valves and direct pilot operated valves do present problems if the pilot pressure is inadequate. In adequate pilot pressure causes improper closure of valve flow passages thus permitting all of the flow or a large portion of the flow to by pass to tank resulting in 'no response' or 'slow response' of the actuator.

Valve compression springs may break due to fatigue in which case the freely floating spool would drift to a new position due to flow forces. This drift may either open a passage or close a passage depending upon the valve configuration thus affecting the function of the machine.

Problems in balanced piston pilot operated relief valves are more frequent than directed operated relief valves. The small orifice in the balanced piston is usually the problem. If it is clogged with dirt the actuator fails to respond.

If the speed of the actuator is varying, that is, if it is not consistent the flow control valve is usually the culprit. Stripping the valve down completely, cleaning all parts in kerosene and re-assembling them after making sure that sliding parts are free to slide would normally correct the problem.

In pilot operated check valves the pilot pressure must be more than 40 per cent of the main line pressure for the check valve to permit return flow. This can be ensured, by inserting a gauge in the main line and in the pilot line. Improper seating of the puppet and broken spring are other causes for failure.

Pressure reducing valves usually have two tapping points, one to record the main line pressure and the other to record the branch line pressure. If the main line pressure is affected it would have an impact on the branch line pressure. So pressure gauges must be inserted at both these points in order to identify the problem.

Problems due to accumulators could be detected by recording the gas pre-charge pressure. This can be carried out by first discharging the oil from the accumulator completely. The nitrogen pre-charge pressure may now be recorded by inserting a test gauge at the tapping point provided on the top of the accumulator.

If this pressure is lower than what is recommended the accumulator must be charged with nitrogen to the correct pressure. If the problem persists the system pressure must be checked by inserting a gauge in the fluid line and by charging the accumulator to the maximum pressure. If the pressure switch is set to correct values the gauge must read the system pressure.

In pre-fill valve circuits the moving platen may close the 'daylight' at normal speed but pressure may not build up or the response time may be too slow. This may be because the main valve is not closing at all or is closing but, not immediately following the closure of 'daylight.' Incorrect setting of the compression spring or the flow control valve as the case may be may cause these problems and must be tackled accordingly. In large hydraulic presses employing 'pre-fill' valves one of the common causes for slow and erratic pressure buildup is air entrapment. It must be ensured that oil occupies the pressure chamber and the pipelines completely, by driving out all air from the system at the highest point in the system. Air bleed valves are normally provided at selected locations especially for this purpose.

Finally, a word about hydraulic oil would be in order although we have discussed this at length in a chapter devoted exclusively for this.

It is estimated that 70 per cent of the cause for hydraulic machine problems can be attributed to defective hydraulic fluid. This should not be a surprise because contamination of oil with water will very quickly cause premature pump failure. Although individual hydraulic component manufacturers recommend a particular grade of hydraulic fluid as suitable for their product, rarely do machine manufactures interpolate these values and recommend the optimum grade of hydraulic oil required for the machine to perform effectively based on the speed of the actuator, working clearances and maximum operating system pressure. Since small and medium scale manufacturers even today, consider preventive maintenance as a luxury they can ill afford, it makes the job of a troubleshooter that much more difficult because he does not know his starting point. For instance if the oil temperature is continuously high for some reason and if the maintenance personnel fail to take note of this fact and take appropriate remedial measures in time, this may itself become the cause for some other system failure. Premature hydraulic seal failure can be attributed to either high oil temperature or contaminated oil. Too high a viscosity would cause additional load on the pump and increased pressure drop resulting in heating up of oil in addition to problems due to cavitations. Too low a viscosity would present leakage problems. Many of the industries buy their oil from retailers instead of from the manufacturer because it is readily available and costs less unaware of the fact that they can be contaminated second hand fluid or at best reclaimed oil. No body knows the viscosity of such oil, or its viscosity index.

9.2 SELECTING ELECTRIC MOTORS FOR INDUSTRIAL FOR HYDRAULIC APPLICATIONS

Three phase squirrel cage A.C. induction motors are the most commonly used motors in industrial hydraulic applications. Among them motors that yield high starting torques and normal break down torque are the most suitable for use on positive displacement pumps. Breakdown torque also called 'pull out torque' is the maximum torque that the motor develops at rated voltage without a sudden drop in speed.

Regarding the type of enclosure both 'open drip proof' (ODP) and 'totally enclosed fan cooled' (TEFC) are equally popular. TEFC design offers greater protection against air borne particles where as ODP design prevents solid particles and liquid droplets from entering the motor.

Cooling fan in case of ODP motors are mounted internally whereas in case of TEFC motors they are mounted externally.

Three types of mountings are available as standard options.

1. Rigid horizontal base mounting.
2. Flange mounting.
3. A combination of rigid horizontal base mounting and face mounting.

This design allows fitting of small pumps using bell-housings direct to the motor face, which has a locating diameter and tapped holes.

Efficiency of the electric motor has a bearing on long-term energy costs. With the spiraling costs of energy it is no longer profitable to accept a low efficiency motor due to its lower initial costs. It is estimated that between two motors of the same class working under similar conditions, you can save an appreciable amount every year in energy costs by selecting a motor with 92 per cent efficiency rating at full load as compared to a motor with 87 per cent efficiency rating.

Motor insulation provides protection for motor windings and with class *E* insulation, which is acceptable in most industrial hydraulic systems; the permitted temperature raise is 75ºC if the room temperature is less than 40ºC. Under these conditions the motor can be operated up to its maximum ratings. If the room temperature raises by say 5 degrees the motor should be de-rated to 95 per cent of its original rating.

India and most other countries have standardized on 50-Hertz frequency except for US where the frequency is 60 Hertz. Provided a constant voltage to frequency ratio is maintained a three-phase electric motor built for 60 Hz can be used with satisfactory results in India. With 50 Hz, frequency is reduced by a factor of 5/6 and so the voltage should be reduced by about the same amount and the 415 volts rating in India is within this limit and therefore adoptable.

However 60 Hz electric motors when adapted to 50 Hz supply run at 5/6 their rated speed. Since speed is proportional to torque, the motor built for 60 Hz but operating on 50 Hz must develop 6/5 or 1.2 times higher torque. This may slightly over load the motor and consequent raise in the operating temperature of the motor. This author has a personal experience of using a hydraulic machine with a 50 Hz electric motor in USA on a 60 Hz supply with very satisfactory results. Since most of the electrical components are rated for both 50/60 Hz it is only the electric motor that one needs to be concerned about. Single-phase 60 Hz motor cannot be used on 50 Hz supply because of wide difference in supply voltage, 110 in USA and 220 in India.

Single-phase motors are not so common in industrial hydraulic applications. This is because three phase motors are more compact, lighter and operate more efficiently, have greater torque values and cost less. If three-phase power is readily available it makes more sense to settle on three phase motors. However where 3-phase supply is not available and where light loads are involved, (HP requirement less than 3) single-phase motors may be chosen. Two most suitable models are the 'Capacitor start-induction run motor' and 'Capacitor start-capacitor run motor'. Generally fractional HP motors are of the first kind and bigger motors are of the second kind.

9.3 SAVE ON ENERGY

There is an enormous scope for the oil hydraulic engineer, to save on energy costs by critically evaluating the input energy required for each and every machine. Unloading valves, Accumulators, intensifiers, variable delivery pumps etc., are all designed with one specific purpose in mind, that is, to save on energy costs. Electric motors have a starting torque rating which is twice that of the running torque. This fact can be used to advantage by the system designer while designing machines like punching presses where the actual shearing or punching operation requires no more than 2-3 seconds. In these applications input HP of the electric motors could be reduced by at least 30 percent over the theoretical values obtained based on pump discharge and maximum system operating pressure. A certain customer had problems with the electric supply company because all his machines together constituted nearly 100 HP while his connected load was much less. He wanted to add a few more machines but was worried about the huge additional power installation costs. He requested me to suggest ways to over come the problem. I suggested to him since he is operating all his machines, mostly hydraulic presses at 50 percent of their rated tonnage he can reduce at least 30 percent of the connected load by installing lower HP electric motors. Also since the cycle time for each machine was 5 minutes during which time the motor was idling, he could use one common power pack for 4 machines. This suggestion was accepted and implemented with huge savings, both long term and short term.

APPENDIX – 1

PROPERTIES OF HYDRAULIC OIL

Compression of Hydraulic Oil at Different Pressures

Pressure (Psi)	Fluid Compression (% of Original Volume)
1000	0.56
2000	1.04
3000	1.47
4000	1.89
5000	2.30
6000	2.67
7000	3.04
8000	3.40
9000	3.75
10000	4.10

Properties of Liquids

	Water	Typical Hydraulic Oil
Density	10.0 kg m^{-3}	850 kg m^{-3}
Bulk Modulus	2.2 × 109 Pa	2.0 × 109 Pa
Specific Heat	4.2 × 103 kg^{-1} K^{-1}	850 J kg^{-1} K^{-1}
Thermal Conduct	0.511 W m^{-1} K^{-1}	0.14 W m^{-1} K^{-1}
Viscosity at 60º C	0.55 × 10^{-3} Nsm^{-2}	9 × 10^{-3} Nsm^{-2}

EFFECT OF PRESSURE ON THE VISCOSITY OF TYPICAL HYDRAULIC OILS

(Temperature = 100 degrees Fahrenheit, Viscosity in Centipoises, Pressures in Pounds per Square inch)

$p = 14.7$	$p = 1,000$		$p = 5,000$		$p = 10,000$	
μ^0	μ	$\beta \times 10^4$	μ	$\beta \times 10^4$	μ	$\beta \times 10$
28.3	33.4	1.65	60.0	1.49	121	1.45
46.4	56.6	1.99	119	1.89	293	1.84
83.1	101	1.99	215	1.90	522	1.84
122	151	2.15	345	2.08	933	2.03
288	351	1.99	714	1.80	1560	1.69
422	515	1.99	1050	1.80	2290	1.69
579	730	2.30	1630	2.16	4070	1.96

APPENDIX – 2

UNIT'S CONVERSION TABLES

Flow Rate

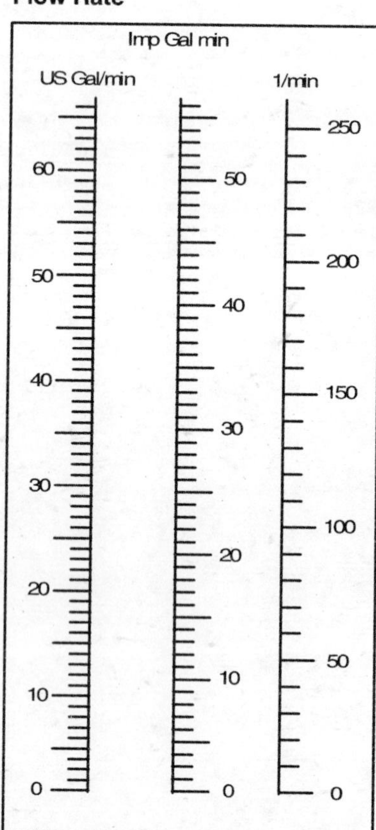

Pressure

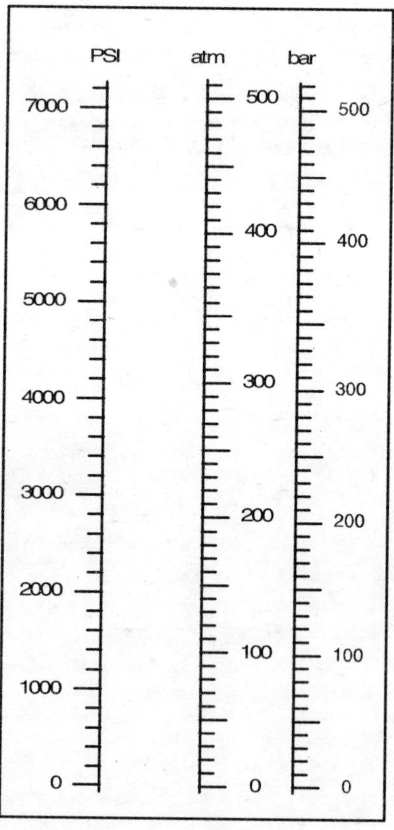

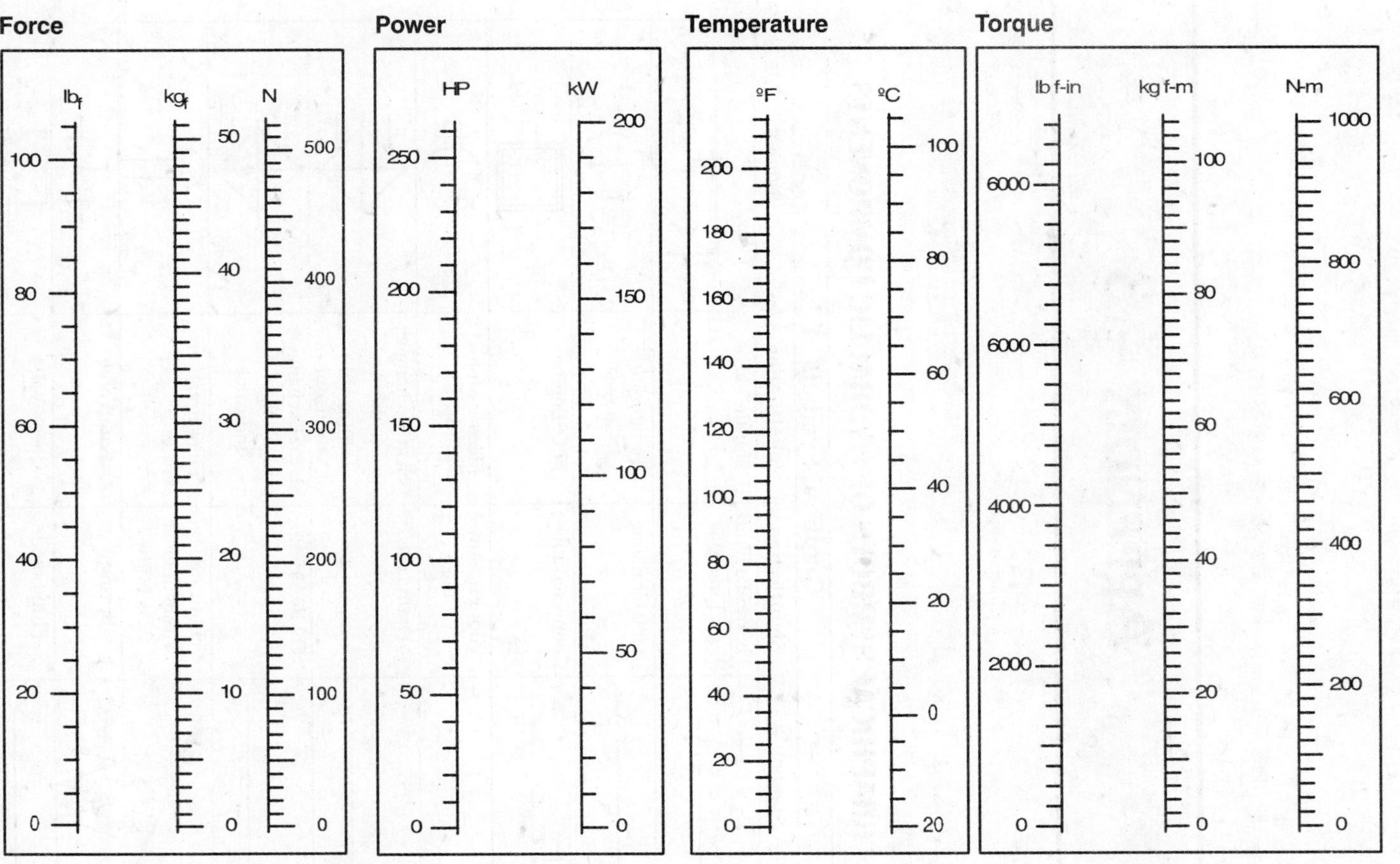
Force
lbf
kgf
N
100
80
60
40
20
0
50
40
30
20
10
0
500
400
300
200
100
0
Power
HP
kW
250
200
150
100
50
0
200
150
100
50
0
Temperature
ºF
ºC
200
180
160
140
120
100
80
60
40
20
0
100
80
60
40
20
0
20
Torque
lb f-in
kg f-m
N-m
6000
6000
4000
2000
0
100
80
60
40
20
0
1000
800
600
400
200
0

APPENDIX – 3

GRAPHICAL SYMBOLS OF HYDRAULIC COMPONENTS

Graphical Symbols - 1			
Symbol	**Signification**	**Signification**	**Symbol**
	Hydraulic Fluid Line	Shuttle Valve	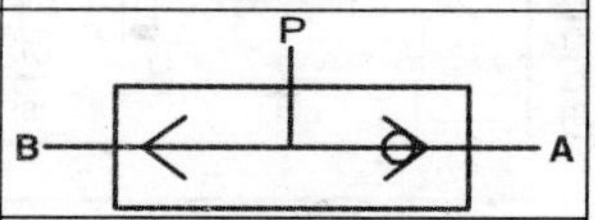 P B A
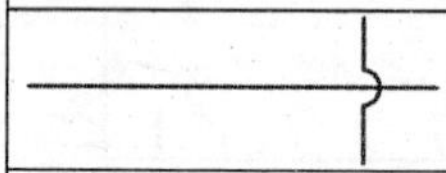	Line Crossing	Temperature Gauge	
	Connected Lines	I.C. Engine	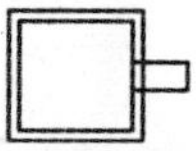
	Pilot Pressure Line	Electric Motor	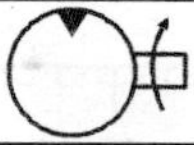
	Drain Line	Heat Exchanger without Temperature Indicator	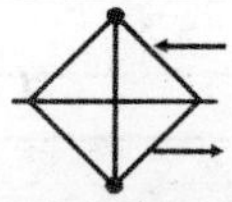
	Oil Reservoir	Heat Exchanger with Temperature Indicator	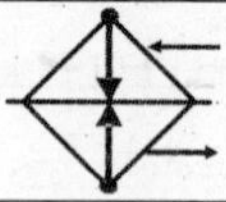
	Indicated Adjustment Possibility	Filler Breather	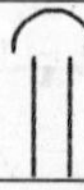
	Spring	Gate Valve	
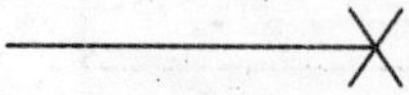	Plug Line	Check Valve	

Symbol	Signification	Signification	Symbol
	Hydraulic HOSE	Hand Pump	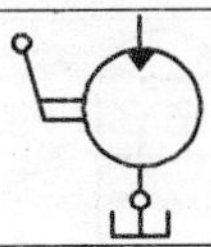
	Quick Connect Coupling	Fixed Displacement Pump	
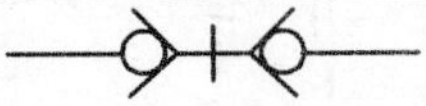	Quick Connect Coupling with Check Valves	Dual Pump	
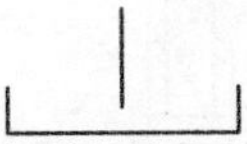	Line to Oil Reservoir		
	Pressure Gauge	Variable Delivery Reversible Pump with Lever Control	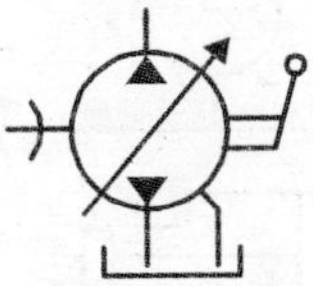
	Inlet Strainer		
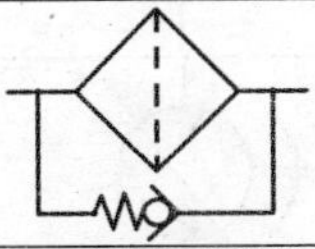	Pressure Line/Return Line Filter	Variable Delivery Reversible Pump with Hand Wheel Control	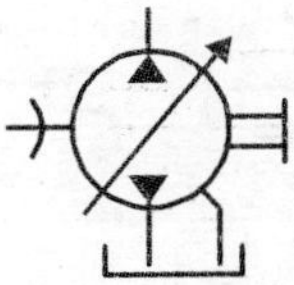
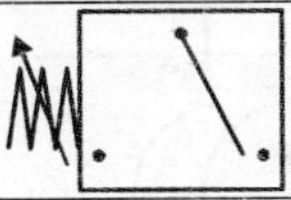	Pressure Switch		

Graphical Symbols - 2			
Symbol	**Signification**	**Signification**	**Symbol**
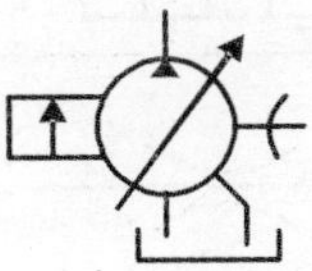	Variable Delivery Pump Pressure Compensation Control	Double Acting Cylinder With Fixed Cushioning at One End	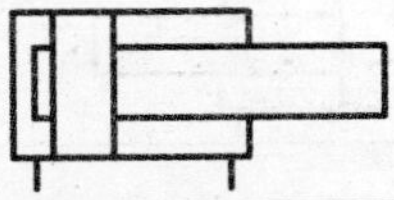
		Double Acting Cylinder with Variable Cushioning at One Ends	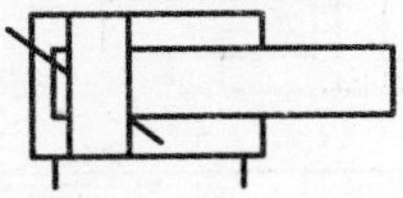
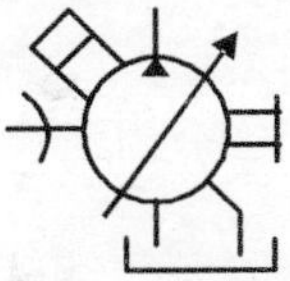	Variable Delivery Pump Pressure Compensated pump with Maximum Adjustable Limit Stop	Double Acting Cylinder With Variable Cushioning at both End	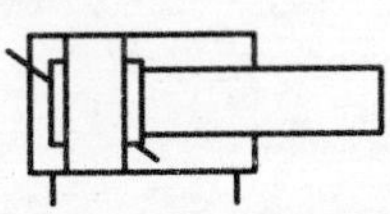
		Duplex Cylinder	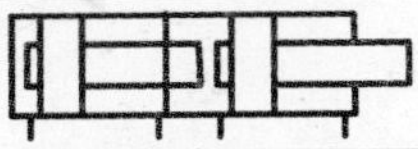

	Weighted Loaded Accumulator	Double Acting Cylinder with Rod at both Ends	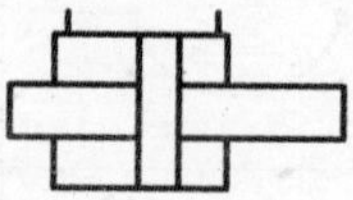
	Spring Loaded Accumulator	Telescoping Cylinder Single Acting	
	Gas Charged Accumulator	Telescoping Cylinder Double Acting	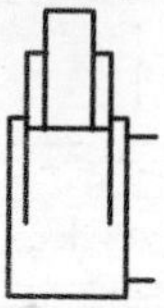
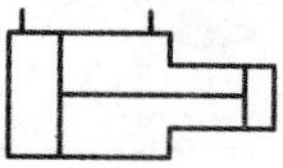	Intensifier Single Acting	Hydraulic Motor Fixed Displacement	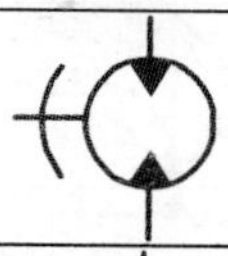
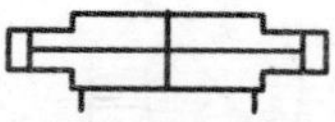	Intensifier Double Acting	Hydraulic Motor Variable Speed	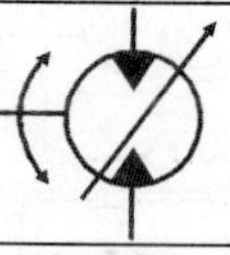
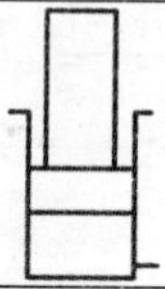	Single Acting Cylinder with Gravity Return	Hydraulic Motor with Limited Angle of Rotation (Torque Generator)	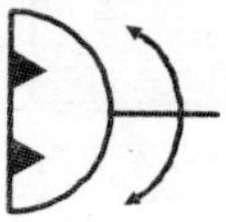
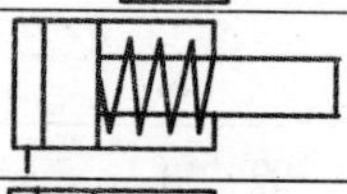	Single Acting Cylinder with Spring Return	Quick Exhaust Valve	
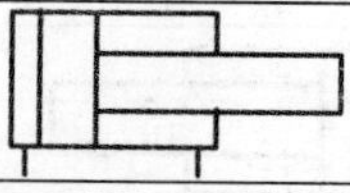	Double Acting Cylinder	Pilot Operated Stop Valve	
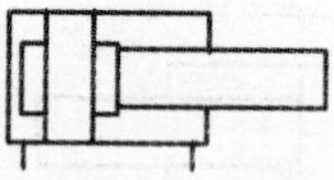	Double Acting Cylinder with Fixed Cushioning at Both Ends	Fixed Orifice Flow Control Valve	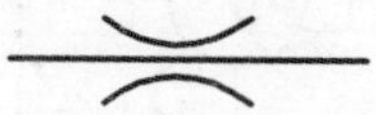

Graphical Symbols - 3

Symbol	Signification	Signification	Symbol
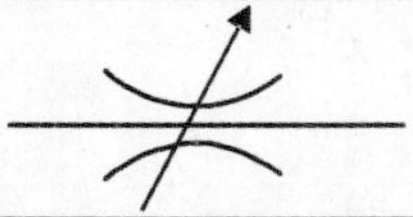	Variable Orifice Flow Control Valve	Two Position 2 Way Directional Spring Offset Control valve Hand Lever Operated	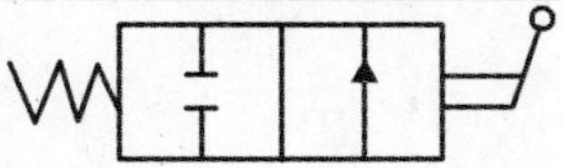
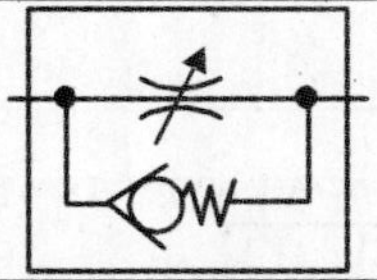	Variable Orifice Flow Control Valve with Integral Check Valve for Free Reserve Flow	Two Position 3 Way Directional Spring Offset Control valve Hand Lever Operated	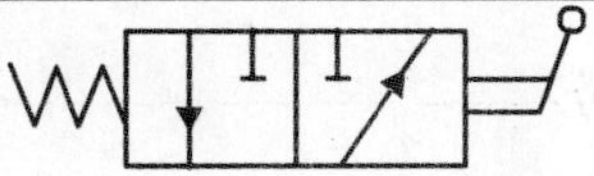
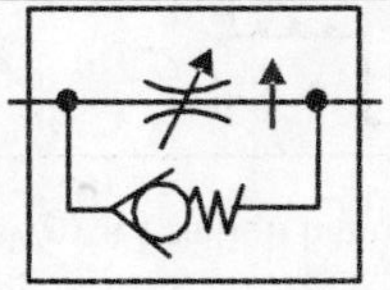	Pressure Compensated Variable Orifice Flow Control Valve with Integral Check Valve for Free Reserve Flow	Solenoid Controlled Pilot Operated Directional Control Valve Simplified Diagram 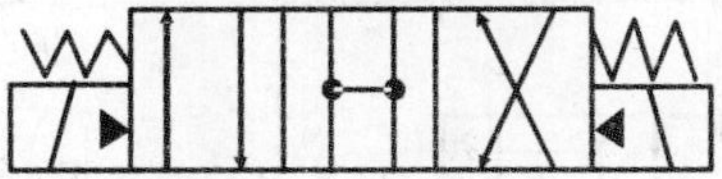 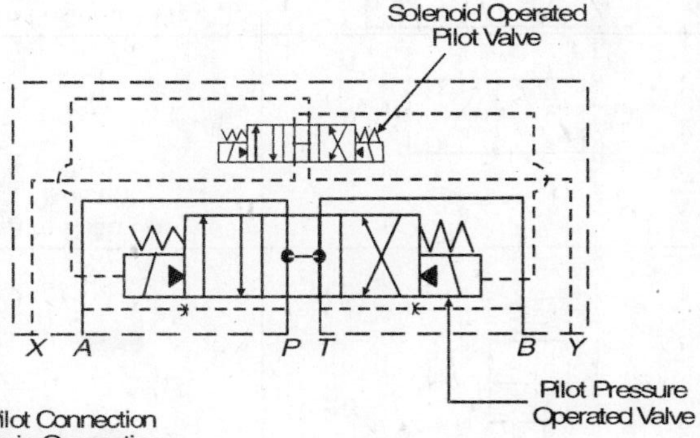 Solenoid Operated Pilot Valve X A P T B Y Pilot Pressure Operated Valve X= Pilot Connection Y= Drain Connection Detailed Diagram	

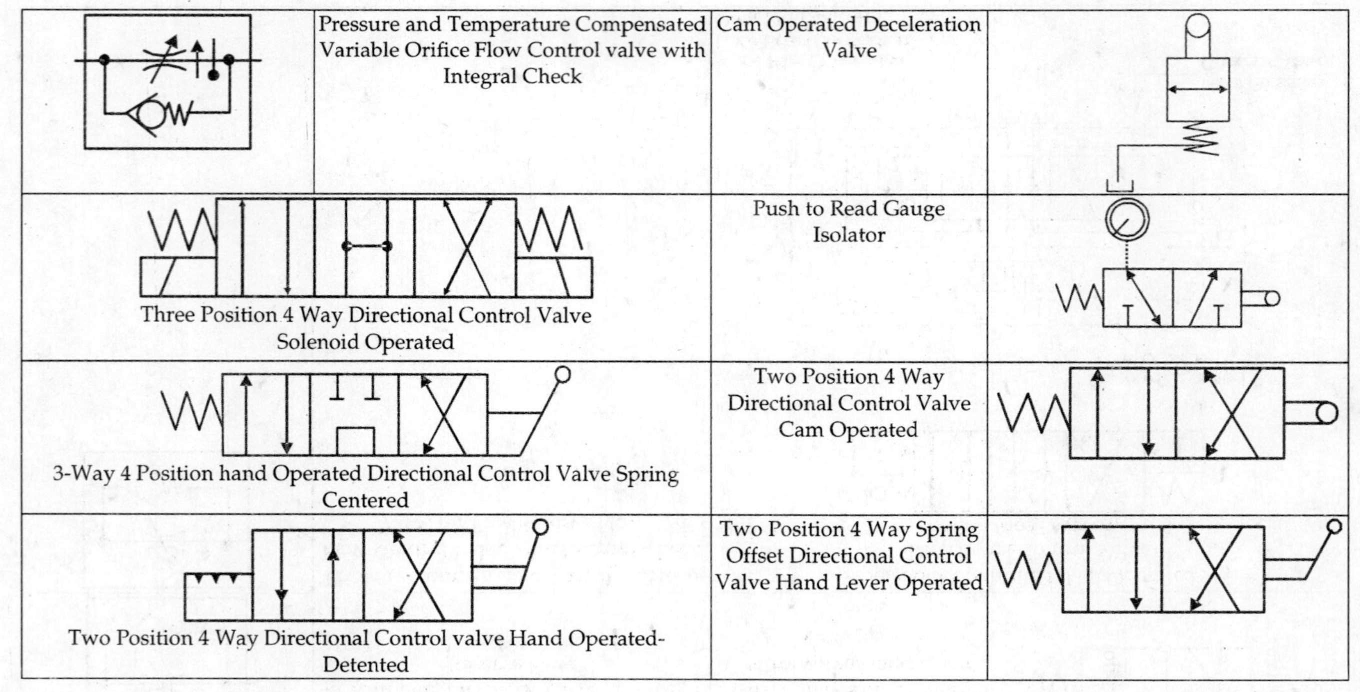

Symbol	Description	Description	Symbol
	Pressure and Temperature Compensated Variable Orifice Flow Control valve with Integral Check	Cam Operated Deceleration Valve	
Three Position 4 Way Directional Control Valve Solenoid Operated		Push to Read Gauge Isolator	
3-Way 4 Position hand Operated Directional Control Valve Spring Centered		Two Position 4 Way Directional Control Valve Cam Operated	
Two Position 4 Way Directional Control valve Hand Operated-Detented		Two Position 4 Way Spring Offset Directional Control Valve Hand Lever Operated	

Graphical Symbols - 4			
Symbol	**Signification**	**Signification**	**Symbol**
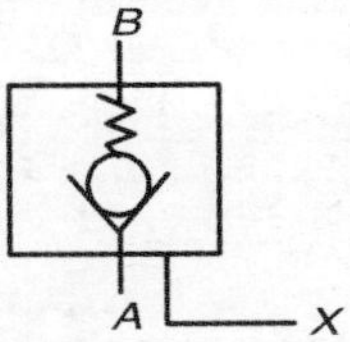	Pilot Operated Check Valve Internal Drain	Counter Balance Valve Internal Pilot Internal Drain	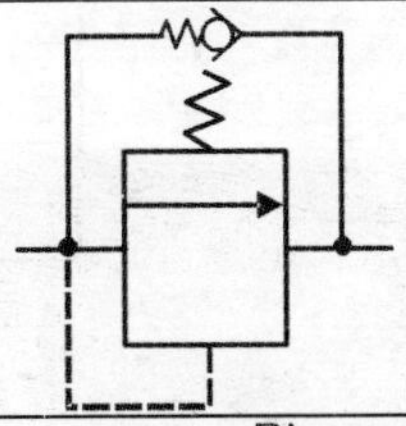
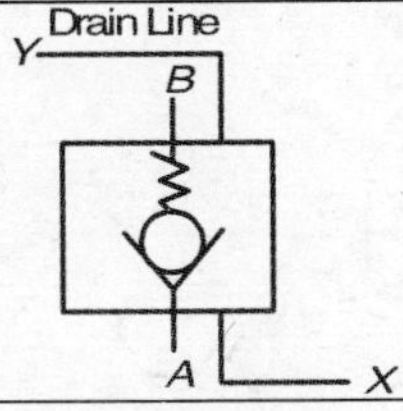	Pilot Operated Check Valve External Drain	Pilot Operated Pressure Relief Valve	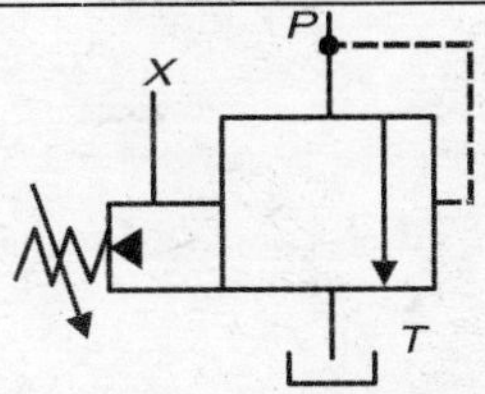
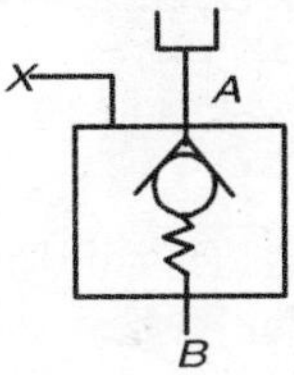	Prefill Valve	Solenoid Controlled Pilot Operated Pressure Relief Valve	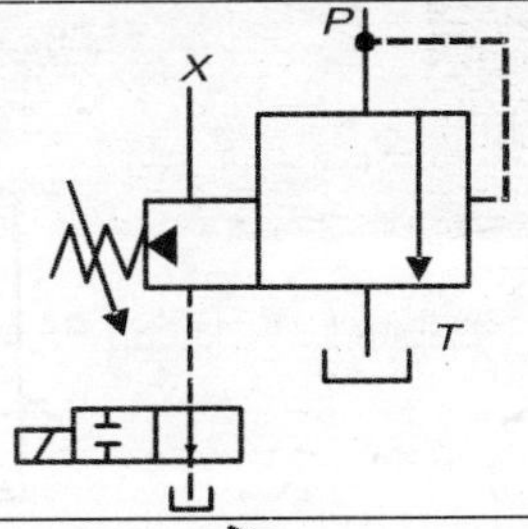
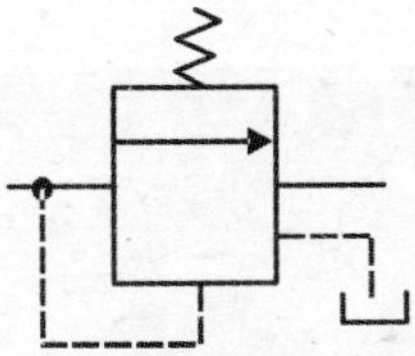	Sequence Valve Internal Pilot External Drain	Pressure Reducing Valve	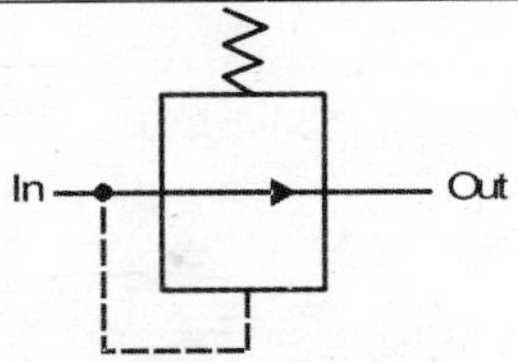
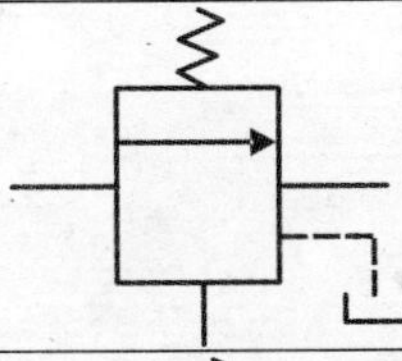	Sequence Valve External Pilot External Drain	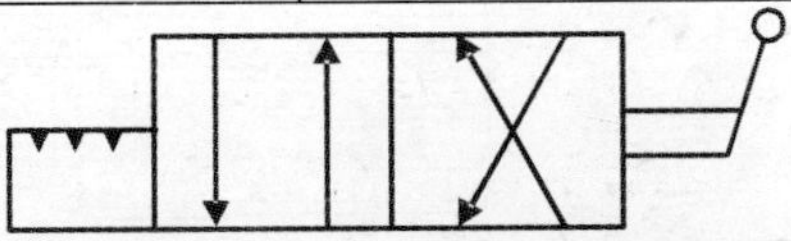 Two Position 4 Way Directional Control Valve Hand Operated - Detented	
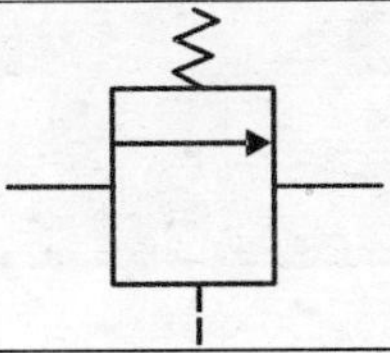	Unloading Relief Valve External Pilot Internal Drain	Direct Operated Pressure Relief Valve	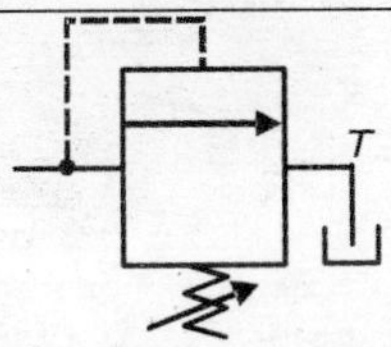

SOME OF THE HYDRAULIC MACHINES DESIGNED BY THE AUTHOR

5-Tonne Workshop Press

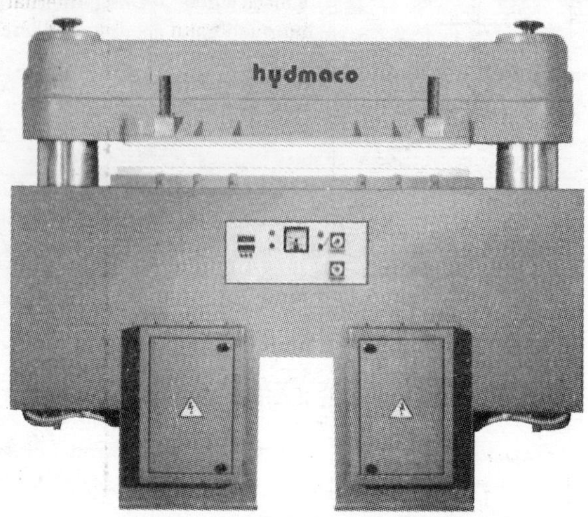

50-Tonne Fully Automatic Electro-Hydraulic Beam Cutting Press

Electro-Hydraulic Swing Arm Die Cutting machine (20 Tonnes)

A Completely Refurbished 600 tonne Double Ram Press Using Prefill Valve System

300-Tonne Hydraulic Press

50-Tonne Hydraulic Press

INDEX